SpringerBriefs in Applied Sciences and Technology

Computational Mechanics

Series Editors

Holm Altenbach, Faculty of Mechanical Engineering,
Otto-von-Guericke-Universität Magdeburg, Magdeburg, Germany

Lucas F. M. da Silva, Department of Mechanical Engineering, Faculty of
Engineering, University of Porto, Porto, Portugal

Andreas Öchsner, Faculty of Mechanical Engineering, Esslingen University of
Applied Sciences, Esslingen, Germany

Reviewers: Prof. Holm Altenbach and Prof. Andreas Öchsner. These SpringerBriefs publish concise summaries of cutting-edge research and practical applications on any subject of computational fluid dynamics, computational solid and structural mechanics, as well as multiphysics.

SpringerBriefs in Computational Mechanics are devoted to the publication of fundamentals and applications within the different classical engineering disciplines as well as in interdisciplinary fields that recently emerged between these areas.

Andreas Öchsner

An Introduction to the Classical Approximation Methods in Applied Mechanics

 Springer

Andreas Öchsner
Faculty of Mechanical and Systems
Engineering
Esslingen University of Applied Sciences
Esslingen, Baden-Württemberg, Germany

ISSN 2191-530X   ISSN 2191-5318 (electronic)
SpringerBriefs in Applied Sciences and Technology
ISSN 2191-5342   ISSN 2191-5350 (electronic)
SpringerBriefs in Computational Mechanics
ISBN 978-3-032-06966-5   ISBN 978-3-032-06967-2 (eBook)
https://doi.org/10.1007/978-3-032-06967-2

This Springer imprint is published by the registered company Springer Nature Switzerland AG
The registered company address is: Gewerbestrasse 11, 6330 Cham, Switzerland

If disposing of this product, please recycle the paper.

Preface

This book is intended as a study aid for the classical approximation methods in undergraduate as well as postgraduate degree courses in engineering at universities. The weighted residual method, i.e. a mathematical procedure, is applied to convert a set of differential equations into equations equal to the number of unknowns based on a given subdivision (e.g. elements, volumes or grid points) of the problem domain. The universal character of this procedure is based on the fact that all classical approximations methods in applied mechanics, i.e., the finite difference method, the finite element method, the finite volume method, and the boundary element method can be derived.

The mechanical theory focuses on the classical one-dimensional structural element of a bar, i.e., a classical tensile member. Focusing on this one-dimensional element reduces the complexity of the mathematical framework and the resulting matrix equations are still possible to be displayed with all components and not only in a symbolic representation. This educational approach proved very useful in different fields of applied mechanics and should facilitate the understanding of the derivations of the various approximation methods. This knowledge is important to extend and implement new features, e.g. constitutive models, in commercial simulation packages.

Esslingen, Germany
July 2025

Andreas Öchsner

Competing Interests The author has no competing interests to declare that are relevant to the content of this manuscript.

Contents

Symbols and Abbreviations

Latin Symbols (Capital Letters)

A	Area
C	Scaler version of elasticity matrix
$\boldsymbol{B}$	Matrix which contains derivatives of interpolation functions
$\boldsymbol{C}$	Elasticity matrix
E	Modulus of elasticity
F	Force
J	Jacobian determinant
$\boldsymbol{K}$	Stiffness matrix
L	Length
N	Internal normal force, interpolation function
$\boldsymbol{N}$	Column matrix of the interpolation functions
$\overline{N}$	Shape function
R	Equivalent nodal force
U	Perimeter
W	Weight function
$W*$	Fundamental solution
$\boldsymbol{W}$	Column matrix of weight functions

Latin Symbols (Small Letters)

$\boldsymbol{b}$	Column matrix of distributed loads
f	Function
f_x	Body force in x-direction
$\boldsymbol{f}$	Column matrix of external forces
g	Function
i	Index number
j	Running index

k	Embedding modulus, running index
n	Number of nodes or grid points, running index
p_x	Distributed force in x-direction
r	Residual
$\boldsymbol{r}$	Residual matrix
t_x	Traction force in x-direction
u	Displacement
$\boldsymbol{u}$	Column matrix of deformations
x	Cartesian coordinate
$\boldsymbol{x}$	Column matrix of Cartesian coordinates

Greek Symbols (Capital Letters)

Ω	Domain

Greek Symbols (Small Letters)

δ	Dirac delta function
$\delta(\ldots)$	Virtual quantity
ε	Normal strain
ξ	Natural coordinate
σ	Normal stress

Mathematical Symbols

$[\ldots]^{\mathrm{T}}$	Transpose
$\dfrac{\mathrm{d}(\ldots)}{\mathrm{d}x}$	First order derivative
$\dfrac{\mathrm{d}^2(\ldots)}{\mathrm{d}x^2}$	Second order derivative
$\mathcal{L}(\ldots)$	Operator symbol
$\mathcal{L}_1(\ldots)$	First oder derivative symbol
$\mathcal{L}_2(\ldots)$	Second oder derivative symbol
$\boldsymbol{\mathcal{L}}$	Matrix of differential operators
$\deg(\ldots)$	Degree of a polynomial
$\Delta\ldots$	Change in a variable
$O(\ldots)$	Order of
sgn	Signum function

Indices, Superscripted

$\ldots^e$ Element

Indices, Subscripted

$\ldots_n$ Degree of derivative

Abbreviations

1D	One-dimensional
BEM	Boundary element method
const.	Constant
FDM	Finite difference method
FEM	Finite element method
FVM	Finite volume method
PDE	Partial differential equation
WRM	Weighted residual method

Chapter 1
Introduction

1.1 Continuum Mechanical Modeling

Continuum mechanical modeling of structural members is based on three fundamental equations [1], i.e., the equilibrium, the kinematics and the constitutive equation, see Fig. 1.1. The equilibrium equation is a measure for the loading of the structure and relates the external loads (forces and in addition moments for bending members) to the internal stresses. The kinematics equation is a measure for the deformation and relates the deformations (displacements and in addition rotations for bending members) to the internal strains. Finally, the constitutive equation or material law (e.g. Hooke's law in the elastic range) relates always the stresses to the strains. Combining these three fundamental equations results, in general, to a system of partial differential equations (PDEs). The final task which remains is the solution of this system of partial differential equations. For very simple problems, it might be possible to obtain the analytical solution of the problem. This analytical solution is, under the made assumptions, exact. However, for more complex problems, one must rely on numerical approximation procedures such as the finite difference method (FDM) [6, 9], the finite element method (FEM) [10, 11], the finite volume method (FVM) [5, 12], or the boundary element method (BEM) [2, 3]. All the methods have in common that they are, in general, no more exact, and the user must guarantee that the approximate solution is as close as possible to the reality. This task requires a solid foundation in engineering courses and a certain level of experience.

The final system of partial differential equations can be expressed for static problems as

$$\mathcal{L}_n u(x) + b(x) = 0 \,, \tag{1.1}$$

where $\mathcal{L}_n$ is a differential operator matrix, which contains nth-order differential operators. For the one-dimensional case, this may simplify, for example, to the scalar operators

© The Author(s), under exclusive license to Springer Nature Switzerland AG 2026
A. Öchsner, *An Introduction to the Classical Approximation Methods in Applied Mechanics*, SpringerBriefs in Computational Mechanics,
https://doi.org/10.1007/978-3-032-06967-2_1

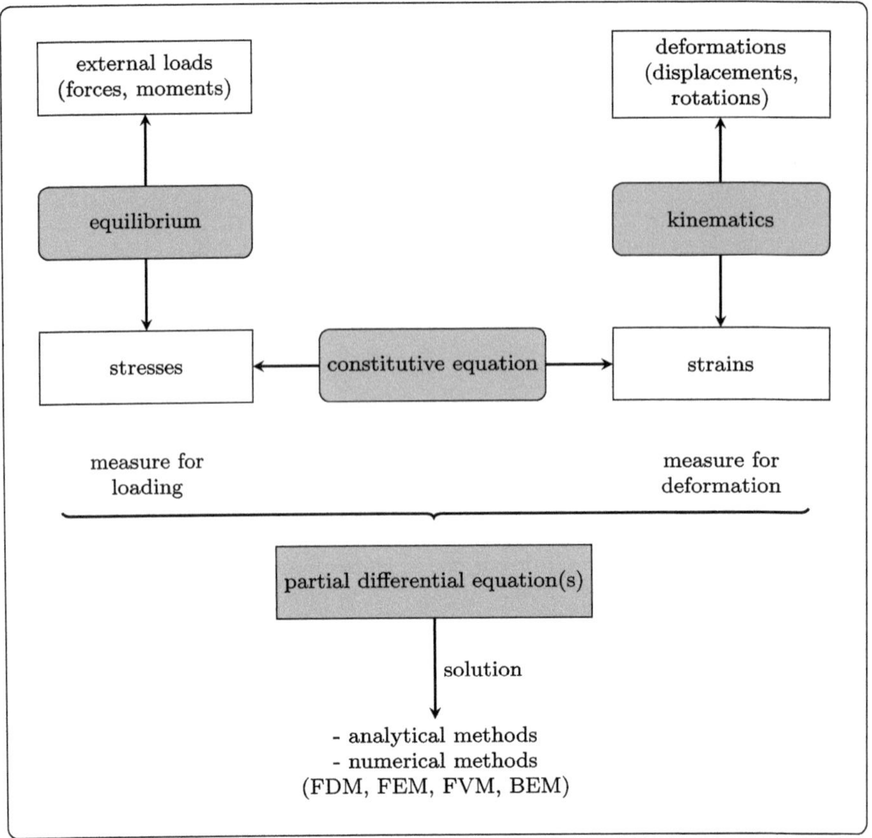

Fig. 1.1 Continuum mechanical modeling

$$\mathcal{L}_1 = \frac{\mathrm{d}(\ldots)}{\mathrm{d}x}\,, \tag{1.2}$$

$$\mathcal{L}_2 = \frac{\mathrm{d}^2(\ldots)}{\mathrm{d}x^2}\,, \tag{1.3}$$

this means the first-order or second-order derivative. The column matrix $\boldsymbol{u}(\boldsymbol{x})$ contains functions (displacements, rotations), which depend on the Cartesian coordinates $\boldsymbol{x}$. The column matrix $\boldsymbol{b}(\boldsymbol{x})$ may also contain a set of functions or simplify to constant values, i.e. $\boldsymbol{b}(\boldsymbol{x}) \rightarrow \boldsymbol{b}$. For the classical mechanical members [8], Eq. (1.1) can be written in the general form

$$\mathcal{L}_{\frac{n}{2}}^{\mathrm{T}} \boldsymbol{C} \mathcal{L}_{\frac{n}{2}} \boldsymbol{u}(\boldsymbol{x}) + \boldsymbol{b} = \boldsymbol{0}\,, \tag{1.4}$$

where $\boldsymbol{C}$ is the so-called elasticity matrix, which is introduced by the constitutive equation.

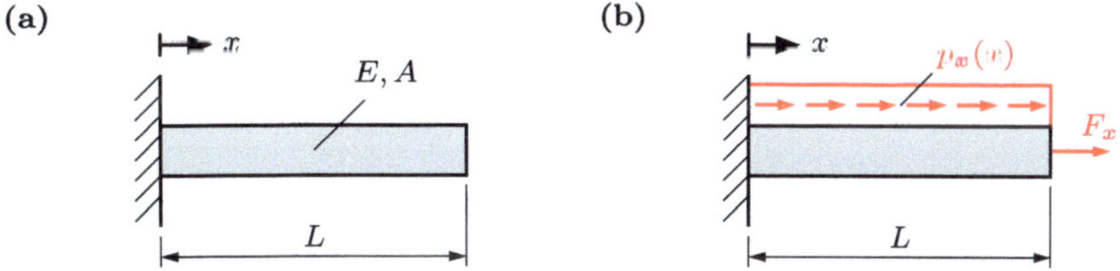

Fig. 1.2 General configuration of an axially loaded bar: **a** geometry and material property; **b** prescribed loads

1.2 Axially Loaded Bar Member

Let us look in the following on a simple mechanical member, i.e. an axially loaded bar, see Fig. 1.2, to illustrate the continuum mechanical modeling approach. In the simplest case which is considered here, the material parameter Young's modulus E and geometric parameter cross-sectional area A are constant. Furthermore, the element length is described by the variable L. On the load side, single forces F_x or distributed loads $p_x(x)$ in units force per unit length can be applied to the structural element. In the case of a body force f_x (unit: force per unit volume), the distributed load takes the form $p_x(x) = f_x(x)A(x)$ where A is the cross-sectional area of the rod. A typical example for a body force would be the dead weight, i.e. the mass under the influence of gravity. In the case of a traction force t_x (unit: force per unit area), the distributed load can be written as $p_x(x) = t_x(x)U(x)$ where $U(x)$ is the perimeter of the cross section. Typical examples are frictional resistance, viscous drag and surface shear.

1.2.1 Kinematics

To derive the strain-displacement relation (kinematics relation), a differential element $\mathrm{d}x$ of such a bar is considered as shown in Fig. 1.3. Under an acting load, this element deforms as indicated in Fig. 1.3b where the initial point at the position x is displaced by u_x and the end point at the position $x + \mathrm{d}x$ is displaced by $u_x + \mathrm{d}u_x$. Thus, the differential element which has a length of $\mathrm{d}x$ in the unloaded state elongates to a length of $\mathrm{d}x + (u_x + \mathrm{d}u_x) - u_x$.

The engineering strain, i.e., the increase in length related to the original length, can be expressed as

$$\varepsilon_x = \frac{(\mathrm{d}x + (u_x + \mathrm{d}u_x) - u_x) - (\mathrm{d}x)}{\mathrm{d}x}, \tag{1.5}$$

or finally as:

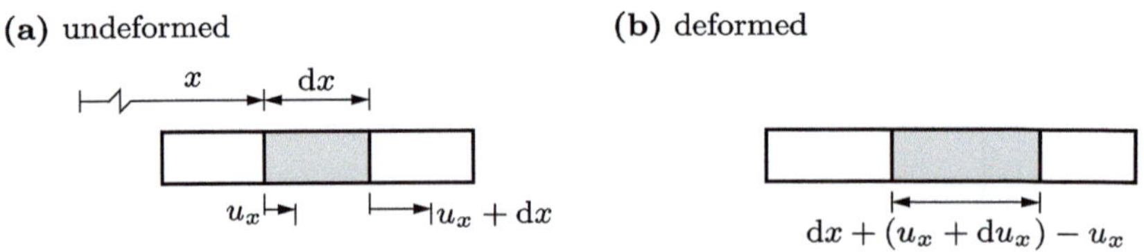

(a) undeformed   **(b) deformed**

Fig. 1.3 Elongation of a differential element of length dx: **a** undeformed configuration; **b** deformed configuration

$$\varepsilon_x(x) = \frac{du_x(x)}{dx}.$$

(1.6)

The last equation is often expressed in a less mathematical way (non-differential) as $\varepsilon_x = \frac{\Delta L}{L}$ where ΔL is the change in length of the entire rod element.

1.2.2 Constitutive Equation

The constitutive equation, i.e., the relation between the stress σ_x and the strain ε_x, is given in its simplest form as Hooke's law

$$\sigma_x(x) = E\varepsilon_x(x),$$

(1.7)

where the Young's modulus E is in the case of linear elasticity a material constant. For the considered rod element, the normal stress and strain is constant over the cross section as shown in Fig. 1.4.

1.2.3 Equilibrium

The equilibrium equation between the external forces and internal reactions can be derived for a differential element of length dx as shown in Fig. 1.5. It is assumed for simplicity that the distributed load p_x and the cross-sectional area A are constant in

(a)   **(b)**

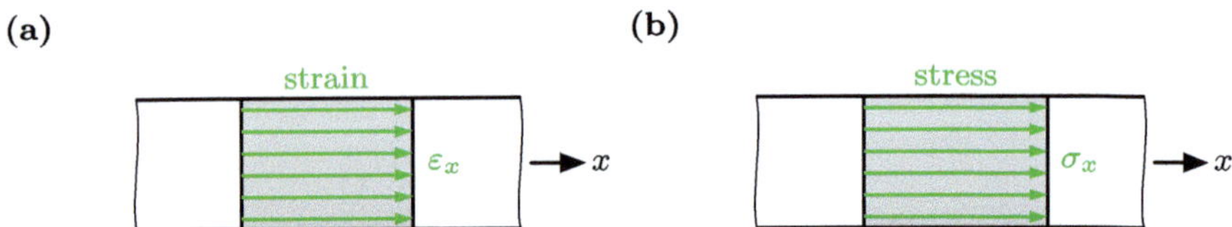

Fig. 1.4 Axially loaded bar: **a** strain and **b** stress distribution

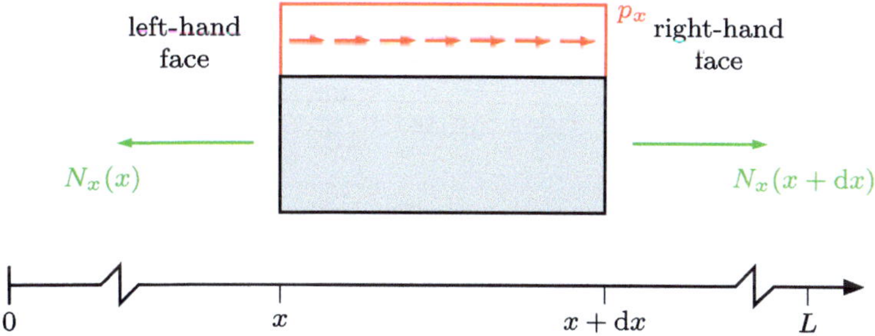

Fig. 1.5 Infinitesimal bar with internal reactions and constant external distributed load

Table 1.1 Fundamental governing equations of a rod for deformation along the x-axis

Expression	Equation
Kinematics	$\varepsilon_x(x) = \dfrac{\mathrm{d}u_x(x)}{\mathrm{d}x}$
Equilibrium	$\dfrac{\mathrm{d}N_x(x)}{\mathrm{d}x} = -p_x(x)$
Constitution	$\sigma_x(x) = E\varepsilon_x(x)$

this figure. The internal reactions N_x are drawn in their positive directions, i.e., at the left-hand face in the negative and at the right-hand face in the positive x-direction.

The force equilibrium in the x-direction for a static configuration requires that

$$- N_x(x) + p_x \mathrm{d}x + N_x(x + \mathrm{d}x) = 0 \tag{1.8}$$

holds. A first-order Taylor's series expansion of the normal force $N_x(x + \mathrm{d}x)$ around point x, i.e.

$$N_x(x + \mathrm{d}x) \approx N_x(x) + \left.\frac{\mathrm{d}N_x}{\mathrm{d}x}\right|_x \mathrm{d}x\,, \tag{1.9}$$

allows to finally express Eq. (1.8) as:

$$\frac{\mathrm{d}N_x(x)}{\mathrm{d}x} = -p_x(x)\,. \tag{1.10}$$

The three fundamental equations to describe the behavior of a rod element are summarized in Table 1.1.

A slightly different derivation of the equilibrium equation is obtained as follows: Eq. (1.8) can be expressed based on the normal stresses as:

$$- \sigma_x(x)A + p_x \mathrm{d}x + \sigma_x(x + \mathrm{d}x)A = 0\,. \tag{1.11}$$

A first-order Taylor's series expansion of the stress $\sigma_x(x + \mathrm{d}x)$ around point x, i.e.

$$\sigma_x(x + \mathrm{d}x) \approx \sigma_x(x) + \left.\frac{\mathrm{d}\sigma_x}{\mathrm{d}x}\right|_x \mathrm{d}x, \tag{1.12}$$

allows to finally express Eq. (1.11) as:

$$\frac{\mathrm{d}\sigma_x(x)}{\mathrm{d}x} + \frac{p_x(x)}{A} = 0. \tag{1.13}$$

The last equation with $\sigma_x = \frac{N_x}{A}$ immediately gives Eq. (1.10).

1.2.4 Differential Equation

To derive the governing partial differential equation, the three fundamental equations given in Table 1.1 must be combined. Introducing the kinematics relation (1.6) into Hooke's law (1.7) gives:

$$\sigma_x(x) = E\frac{\mathrm{d}u_x}{\mathrm{d}x}. \tag{1.14}$$

Considering in the last equation that a normal stress is defined as an acting force N_x over a cross-sectional area A:

$$\frac{N_x}{A} = E\frac{\mathrm{d}u_x}{\mathrm{d}x}. \tag{1.15}$$

The last equation can be differentiated with respect to the x-coordinate to give:

$$\frac{\mathrm{d}N_x}{\mathrm{d}x} = \frac{\mathrm{d}}{\mathrm{d}x}\left(EA\frac{\mathrm{d}u_x}{\mathrm{d}x}\right), \tag{1.16}$$

where the derivative of the normal force can be replaced by the equilibrium equation (1.10) to obtain in the general case:

$$\frac{\mathrm{d}}{\mathrm{d}x}\left(E(x)A(x)\frac{\mathrm{d}u_x(x)}{\mathrm{d}x}\right) = -p_x(x). \tag{1.17}$$

The general case in the formulation with the internal normal force distribution reads:

$$E(x)A(x)\frac{\mathrm{d}u_x(x)}{\mathrm{d}x} = N_x(x). \tag{1.18}$$

Table 1.2 Different formulations of the partial differential equation for a rod (x-axis: right facing)

Configuration	Partial differential equation
E, A	$EA\dfrac{\mathrm{d}^2 u_x}{\mathrm{d}x^2} = 0$
$E(x), A(x)$	$\dfrac{\mathrm{d}}{\mathrm{d}x}\left(E(x)A(x)\dfrac{\mathrm{d}u_x}{\mathrm{d}x}\right) = 0$
$p_x(x)$	$EA\dfrac{\mathrm{d}^2 u_x}{\mathrm{d}x^2} = -p_x(x)$
$k(x)$	$EA\dfrac{\mathrm{d}^2 u_x}{\mathrm{d}x^2} = k(x)u_x$

Thus, to obtain the displacement field $u_x(x)$, one may start from Eq. (1.17) or from Eq. (1.18). The first approach requires to state the distribution of the distributed load $p_x(x)$ while for the second approach one requires the internal normal force distribution $N_x(x)$.

If the axial tensile stiffness EA is constant, the formulation (1.17) can be simplified to:

$$EA\frac{\mathrm{d}^2 u_x(x)}{\mathrm{d}x^2} = -p_x(x)\,. \tag{1.19}$$

Some common formulations of the governing partial differential equation are collected in Table 1.2. It should be noted here that some of the different cases given in Table 1.2 can be combined. The last case in Table 1.2 refers to the case of elastic embedding of a rod where the embedding modulus k has the unit of force per unit area. Analytical solutions for different loading and support conditions can be found, for example, in [7].

If we replace the common formulation of the first order derivative, i.e. $\frac{\mathrm{d}(\dots)}{\mathrm{d}x}$, by a formal operator symbol, i.e. $\mathcal{L}_1(\dots)$, the basic equations can be stated in a more formal way as given in Table 1.3. Such a formulation is advantageous in the two- and three-dimensional cases [1]. It should be noted here that the transposed ('T'), i.e., $\mathcal{L}_1^{\mathrm{T}} = \left(\frac{\mathrm{d}(\dots)}{\mathrm{d}x}\right)^{\mathrm{T}}$, is only used to show similar structures of the equations in the two- and three-dimensional case [8].

Looking at the general formulation provided in Eq. (1.4), it is easy to see the similarity in the notation of the differential equations:

Table 1.3 Different formulations of the basic equations for a rod (x-axis along the principal rod axis). E: Young's modulus; A: cross-sectional area; p_x: length-specific distributed normal load; $\mathcal{L}_1 = \frac{\mathrm{d}(\dots)}{\mathrm{d}x}$: first-order derivative; b: volume-specific distributed normal load

Specific formulation	General formulation
Kinematics $$\varepsilon_x(x) = \frac{\mathrm{d}u_x(x)}{\mathrm{d}x}$$	$$\varepsilon_x(x) = \mathcal{L}_1\left(u_x(x)\right)$$
Constitution $$\sigma_x(x) = E\varepsilon_x(x)$$	$$\sigma_x(x) = C\varepsilon_x(x)$$
Equilibrium $$\frac{\mathrm{d}\sigma_x(x)}{\mathrm{d}x} + \frac{p_x(x)}{A} = 0$$	$$\mathcal{L}_1^{\mathrm{T}}\left(\sigma_x(x)\right) + b = 0$$
PDE ($A = \text{const.}$) $$\frac{\mathrm{d}}{\mathrm{d}x}\left(E(x)\frac{\mathrm{d}u_x}{\mathrm{d}x}\right) + \frac{p_x(x)}{A} = 0$$	$$\mathcal{L}_1^{\mathrm{T}}\left(C\mathcal{L}_1\left(u_x(x)\right)\right) + b = 0$$ or $\mathcal{L}_1^{\mathrm{T}}\left(EA\mathcal{L}_1\left(u_x(x)\right)\right) + p_x = 0$

$$\mathcal{L}_{\frac{n}{2}}^{\mathrm{T}}\, C \mathcal{L}_{\frac{n}{2}}\, u(x) + b = 0 \tag{1.20}$$

$$\downarrow$$

$$\mathcal{L}_1^{\mathrm{T}}\left(EA\mathcal{L}_1\left(u_x(x)\right)\right) + p_x = 0\,. \tag{1.21}$$

1.3 Weighted Residual Method

The weighted residual method (WRM) is a mathematical procedure to convert a set of differential equations into equations equal to the number of unknowns based on a given subdivision (e.g. elements, volumes or grid points) of the problem domain. The universal character of this procedure is based on the fact that all classical approximations methods, i.e., the finite difference method, the finite element method, the finite volume method, and the boundary element method can be derived, see Fig. 1.6. Let us briefly explain the approach based on the general formulation.

Starting point is the so-called *strong formulation*, i.e., the partial differential equation as given in Eq. (1.4):

$$\mathcal{L}_{\frac{n}{2}}^{\mathrm{T}}\, C \mathcal{L}_{\frac{n}{2}}\, u(x) + b = 0\,. \tag{1.22}$$

This formulation contains in general the approximate solution and is in the scope of approximation methods no longer fulfilled. As a result, a *residual* matrix r is obtained:

$$r = \mathcal{L}_{\frac{n}{2}}^{\mathrm{T}}\, C \mathcal{L}_{\frac{n}{2}}\, u(x) + b \neq 0\,. \tag{1.23}$$

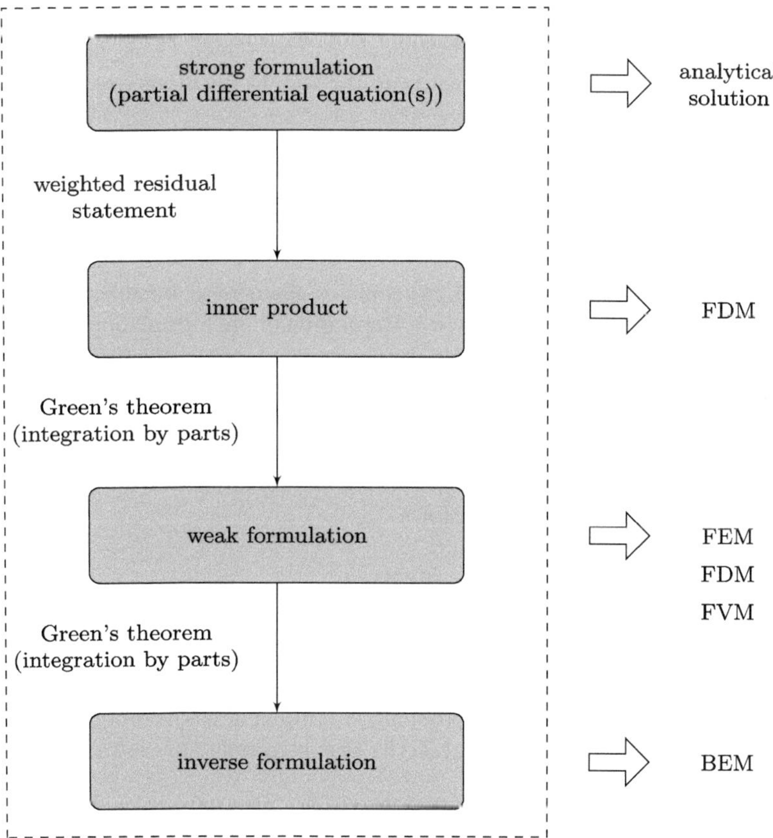

Fig. 1.6 Some classical approximation methods in the context of the weighted residual method

Alternatively, it is now requested based on an integral statement, the so-called *inner product*, that the differential equation is fulfilled in average sense in a certain domain Ω:

$$\int_{\Omega} \boldsymbol{W}^{\mathrm{T}}(\boldsymbol{x}) \left(\mathcal{L}_{\frac{n}{2}}^{\mathrm{T}} \boldsymbol{C} \mathcal{L}_{\frac{n}{2}} \boldsymbol{u}(\boldsymbol{x}) + \boldsymbol{b} \right) \mathrm{d}\Omega \overset{!}{=} 0 \,, \tag{1.24}$$

where $\boldsymbol{W}^{\mathrm{T}}(\boldsymbol{x})$ is the column matrix of weight functions which distribute the error in the considered domain Ω. Shifting one of the differential operator matrices to the weight functions under application of the Green-Gauss theorem yields the co-called *weak formulation* as:

$$\int_{\Omega} \left(\mathcal{L}_{\frac{n}{2}} \boldsymbol{W}(\boldsymbol{x}) \right)^{\mathrm{T}} \boldsymbol{C} \left(\mathcal{L}_{\frac{n}{2}} \boldsymbol{u}(\boldsymbol{x}) \right) \mathrm{d}\Omega = \dots \,, \tag{1.25}$$

where the right-hand side of Eq. (1.25) contains boundary expressions and a domain integral with the matrix $\boldsymbol{b}$. A further application of the Green-Gauss theorem allows to shift also the second operator matrix to the weight functions and the so-called *inverse formulation* is obtained as:

$$\int_{\Omega} \left(\mathcal{L}_{\frac{n}{2}}^{\mathrm{T}} \boldsymbol{C} \mathcal{L}_{\frac{n}{2}} \boldsymbol{W}(\boldsymbol{x}) \right)^{\mathrm{T}} \boldsymbol{u}(\boldsymbol{x}) \mathrm{d}\Omega = \dots , \tag{1.26}$$

where the right-hand side of Eq. (1.26) contains again some boundary expressions and a domain integral with the matrix $\boldsymbol{b}$. Depending of the formulation, i.e. from the strong to the inverse formulation, and the particular choice of the weight functions, all the classical approximation methods commonly used in applied mechanics can be derived.

Let us now explain the approach in detail at the example of a bar element as shown in Fig. 1.2. Let us start from the partial differential equation, i.e., the *strong form* of the problem, for a tensile bar in the form

$$\frac{\mathrm{d}}{\mathrm{d}x} \left(E(x) A(x) \frac{\mathrm{d}u_x^0(x)}{\mathrm{d}x} \right) + p_x(x) = 0 , \tag{1.27}$$

where $u^0(x)$ represents the *exact* solution of the problem. The last equation, which contains the exact solution of the problem, is fulfilled at any position x of the bar. Replacing the exact solution in Eq. (1.27) by an approximate solution $u(x)$, a residual r is obtained:

$$r(x) = \frac{\mathrm{d}}{\mathrm{d}x} \left(E(x) A(x) \frac{\mathrm{d}u_x(x)}{\mathrm{d}x} \right) + p_x(x) \neq 0 . \tag{1.28}$$

As a consequence of the introduction of the approximate solution $u(x)$, it is in general no longer possible to satisfy the differential equation at each position x of the bar. It is alternatively requested in the following that the differential equation is fulfilled over a certain range (and no longer at each location x) and the following integral statement is obtained

$$\int_0^L W(x) \left(\frac{\mathrm{d}}{\mathrm{d}x} \left(E(x) A(x) \frac{\mathrm{d}u_x(x)}{\mathrm{d}x} \right) + p_x(x) \right) \mathrm{d}x \overset{!}{=} 0 , \tag{1.29}$$

which is called the *inner product*. The function $W(x)$ in Eq. (1.29) is called the weight function which distributes the error or the residual in the considered domain. Alternatively, the weight function is sometimes called the test function. Let us split Eq. (1.29) in two integrals:

$$\int_0^L W(x) \left(\frac{\mathrm{d}}{\mathrm{d}x} \left(E(x) A(x) \frac{\mathrm{d}u_x(x)}{\mathrm{d}x} \right) \right) \mathrm{d}x + \int_0^L W(x) p_x(x) \mathrm{d}x \overset{!}{=} 0 . \qquad (1.30)$$

Integrating by parts[1] of the first integral in Eq. (1.30) gives

$$\int_0^L \underbrace{W(x)}_{f} \underbrace{\left(\frac{\mathrm{d}}{\mathrm{d}x} \left(E(x) A(x) \frac{\mathrm{d}u_x(x)}{\mathrm{d}x} \right) \right)}_{g'} \mathrm{d}x = \left[\underbrace{W(x)}_{f} \underbrace{E(x) A(x) \frac{\mathrm{d}u_x(x)}{\mathrm{d}x}}_{g} \right]_0^L$$

$$- \int_0^L \underbrace{\frac{\mathrm{d}W(x)}{\mathrm{d}x}}_{f'} \underbrace{E(x) A(x) \frac{\mathrm{d}u_x(x)}{\mathrm{d}x}}_{g} \mathrm{d}x , \qquad (1.31)$$

which can be introduced in Eq. (1.30) to give the so-called *weak formulation* of the problem as:

$$\int_0^L \frac{\mathrm{d}W(x)}{\mathrm{d}x} E(x) A(x) \frac{\mathrm{d}u_x(x)}{\mathrm{d}x} \mathrm{d}x = \left[W(x) E(x) A(x) \frac{\mathrm{d}u_x(x)}{\mathrm{d}x} \right]_0^L + \int_0^L W(x) p_x(x) \mathrm{d}x .$$

$$(1.32)$$

Looking at the weak formulation, it can be seen that the integration by parts shifted one derivative from the approximate solution to the weight function and a symmetrical formulation with respect to the derivatives is obtained. Integration by parts of the integral on the left hand side in the weak form according to Eq. (1.32) gives

$$\int_0^L \underbrace{\frac{\mathrm{d}W(x)}{\mathrm{d}x}}_{f} \underbrace{E(x) A(x) \frac{\mathrm{d}u_x(x)}{\mathrm{d}x}}_{g'} \mathrm{d}x = \left[\underbrace{\frac{\mathrm{d}W(x)}{\mathrm{d}x}}_{f} \underbrace{E(x) A(x) u_x(x)}_{g} \right]_0^L \qquad (1.33)$$

$$- \int_0^L \underbrace{\frac{\mathrm{d}^2 W(x)}{\mathrm{d}x^2}}_{f'} \underbrace{E(x) A(x) u_x(x)}_{g} \mathrm{d}x , \qquad (1.34)$$

which can be introduced in Eq. (1.32) to give the so-called *inverse formulation* of the problem as:

---

[1] A common representation of integration by parts of two functions $f(x)$ and $g(x)$ is: $\int fg' \mathrm{d}x = fg - \int f'g \, \mathrm{d}x$.

Table 1.4 Summary: Different formulations for a tensile bar in the context of the weighted residual method

Strong formulation (PDE)

$$\frac{\mathrm{d}}{\mathrm{d}x}\left(E(x)A(x)\frac{\mathrm{d}u_x^0(x)}{\mathrm{d}x}\right) + p_x(x) = 0$$

Inner product

$$\int\limits_0^L W(x)\left(\frac{\mathrm{d}}{\mathrm{d}x}\left(E(x)A(x)\frac{\mathrm{d}u_x(x)}{\mathrm{d}x}\right) + p_x(x)\right)\mathrm{d}x \overset{!}{=} 0$$

Weak formulation

$$\int\limits_0^L \frac{\mathrm{d}W(x)}{\mathrm{d}x}E(x)A(x)\frac{\mathrm{d}u_x(x)}{\mathrm{d}x}\mathrm{d}x = \left[W(x)E(x)A(x)\frac{\mathrm{d}u_x(x)}{\mathrm{d}x}\right]_0^L + \int\limits_0^L W(x)p_x(x)\mathrm{d}x$$

Inverse formulation

$$\int\limits_0^L \frac{\mathrm{d}^2W(x)}{\mathrm{d}x^2}E(x)A(x)u_x(x)\mathrm{d}x = \left[\frac{\mathrm{d}W(x)}{\mathrm{d}x}E(x)A(x)u_x(x)\right]_0^L - \left[W(x)E(x)A(x)\frac{\mathrm{d}u_x(x)}{\mathrm{d}x}\right]_0^L$$

$$- \int\limits_0^L W(x)p_x(x)\mathrm{d}x$$

$$\int\limits_0^L \frac{\mathrm{d}^2W(x)}{\mathrm{d}x^2}E(x)A(x)u_x(x)\mathrm{d}x = \left[\frac{\mathrm{d}W(x)}{\mathrm{d}x}E(x)A(x)u_x(x)\right]_0^L$$

$$- \left[W(x)E(x)A(x)\frac{\mathrm{d}u_x(x)}{\mathrm{d}x}\right]_0^L \tag{1.35}$$

$$- \int\limits_0^L W(x)p_x(x)\mathrm{d}x\,.$$

The characteristic of the inverse formulation is that the differential operator is completely shifted from $u_x(x)$ to the weight function $W(x)$. The different formulations in the scope of the weighted residual method are summarized in Table 1.4 for a tensile bar.

References

1. H. Altenbach, A. Öchsner (eds.), *Encyclopedia of Continuum Mechanics* (Springer, Berlin, 2020)
2. C.A. Brebbia, J.C.F. Telles, L.C. Wrobel, *Boundary Element Techniques: Theory and Applications in Engineering* (Springer, Berlin, 1984)
3. L. Gaul, M. Kögl, M. Wagner, *Boundary Element Methods for Engineers and Scientists: An Introductory Course with Advanced Topics* (Springer, Berlin, 2003)

4. Z. Javanbakht, A. Öchsner, *Computational Statics Revision Course* (Springer, Cham, 2018)
5. R.J. LeVeque, *Finite Volume Methods for Hyperbolic Problems* (Cambridge University Press, Cambridge, 2004)
6. R.J. LeVeque, *Finite Difference Methods for Ordinary and Partial Differential Equations: Steady-State and Time-Dependent Problems* (Society for Industrial and Applied Mathematics SIAM, Philadelphia, 2007)
7. A. Öchsner, *Elasto-Plasticity of Frame Structure Elements: Modeling and Simulation of Rods and Beams* (Springer-Verlag, Berlin, 2014)
8. A. Öchsner, *Partial Differential Equations of Classical Structural Members: A Consistent Approach* (Springer, Cham, 2020)
9. A. Öchsner, *Structural Mechanics with a Pen: A Guide to Solve Finite Difference Problems* (Springer, Cham, 2021)
10. A. Öchsner, *Computational Statics and Dynamics: An Introduction Based on the Finite Element Method* (Springer, Singapore, 2023)
11. M. Merkel, A. Öchsner, *One-Dimensional Finite Elements: An Introduction To The Method* (Springer Vieweg, Berlin, 2023)
12. R. Petrova, *Finite Volume Method: Powerful Means of Engineering Design* (InTech, Rijeka, 2012)

Chapter 2
Finite Difference Method

2.1 Derivation Based on the Strong Formulation

Starting point for the derivation of the finite difference formulation is the *strong formulation* as given in Eq. (1.19), i.e. the special case for constant[1] tensile stiffness:

$$EA\frac{\mathrm{d}^2 u_x(x)}{\mathrm{d}x^2} = -p_x(x)\,. \tag{2.1}$$

As a first step to solve the differential equation for a one-dimensional problem, it is assumed that the numerical solution $u(x)$ is to be determined only at a set of n points within the domain $x \in [0, L]$ including the ends, cf. Fig. 2.1.

These n nodes or grid points will be used to derive approximations to the derivatives of the function $u(x)$ based on Taylor's series expansions. To simplify the approach, let us assume in the following that these n nodes are equally spaced, at a distance equal to[2] $\Delta x = \frac{L}{n-1}$.

A typical inner node of the domain is denoted in the following by i and the two neighboring nodes are called $i - 1$ on the left-hand and $i + 1$ on the right-hand side. For sufficient smooth functions $u(x)$, a Taylor's series expansion around node i gives:

$$u_{i+1} = u_i + \left(\frac{\mathrm{d}u}{\mathrm{d}x}\right)_i \Delta x + \left(\frac{\mathrm{d}^2 u}{\mathrm{d}x^2}\right)_i \frac{\Delta x^2}{2} + \cdots + \left(\frac{\mathrm{d}^k u}{\mathrm{d}x^k}\right)_i \frac{\Delta x^k}{k!}\,, \tag{2.2}$$

$$u_{i-1} = u_i - \left(\frac{\mathrm{d}u}{\mathrm{d}x}\right)_i \Delta x + \left(\frac{\mathrm{d}^2 u}{\mathrm{d}x^2}\right)_i \frac{\Delta x^2}{2} - \cdots + \left(\frac{\mathrm{d}^k u}{\mathrm{d}x^k}\right)_i \frac{\Delta x^k}{k!}\,. \tag{2.3}$$

---

[1] The derivation for varying material and geometry parameters is presented, for example, in [6].

[2] It should be noted here that the first node in this derivation is denoted by '1' and not as in some references as '0'.

© The Author(s), under exclusive license to Springer Nature Switzerland AG 2026
A. Öchsner, *An Introduction to the Classical Approximation Methods
in Applied Mechanics*, SpringerBriefs in Computational Mechanics,
https://doi.org/10.1007/978-3-032-06967-2_2

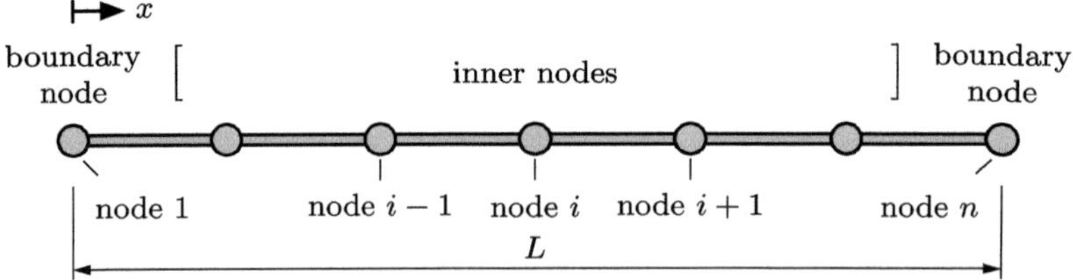

Fig. 2.1 Finite difference model of a one-dimensional problem

The infinite series of Eqs. (2.2) and (2.3) are truncated for practical use after a certain number of terms. As a result of this approximation, the so-called truncation errors occurs. Summing up the expression of Eqs. (2.2) and (2.3) gives

$$u_{i+1} + u_{i-1} = 2u_i + \left(\frac{d^2u}{dx^2}\right)_i \Delta x^2 + \frac{1}{12}\left(\frac{d^4u}{dx^4}\right)_i \Delta x^4 + \cdots, \qquad (2.4)$$

or rearranged for the second order derivative:

$$\left(\frac{d^2u}{dx^2}\right)_i = \frac{u_{i+1} - 2u_i + u_{i-1}}{\Delta x^2} \underbrace{- \frac{1}{12}\left(\frac{d^4u}{dx^4}\right)_i \Delta x^2 - \cdots}_{O(\Delta x^2)}. \qquad (2.5)$$

The symbol 'O' in Eq. (2.5) reads 'order of' and states that if the second order derivative of $u(x)$ is approximated by the first expression on the right-hand side of Eq. (2.5), then the truncation error is of order of Δx^2. This approximation is a second order accurate approximation because of the truncated terms and is called the centered difference scheme. In a similar way, other derivatives can be derived from Eqs. (2.2) and (2.3). Subtracting of Eq. (2.3) from (2.2) gives

$$u_{i+1} - u_{i-1} = 2\left(\frac{du}{dx}\right)_i \Delta x + \frac{1}{3}\left(\frac{d^3u}{dx^3}\right)_i \Delta x^3 + \cdots \qquad (2.6)$$

or rearranged for the first order derivative

$$\left(\frac{du}{dx}\right)_i = \frac{u_{i+1} - u_{i-1}}{2\Delta x} \underbrace{- \frac{1}{6}\left(\frac{d^3u}{dx^3}\right)_i \Delta x^2 - \cdots}_{O(\Delta x^2)}. \qquad (2.7)$$

This last approximation of the first order derivative is called centered difference or centered Euler and the truncation error is of order Δx^2. Rearranging Eq. (2.2), i.e.

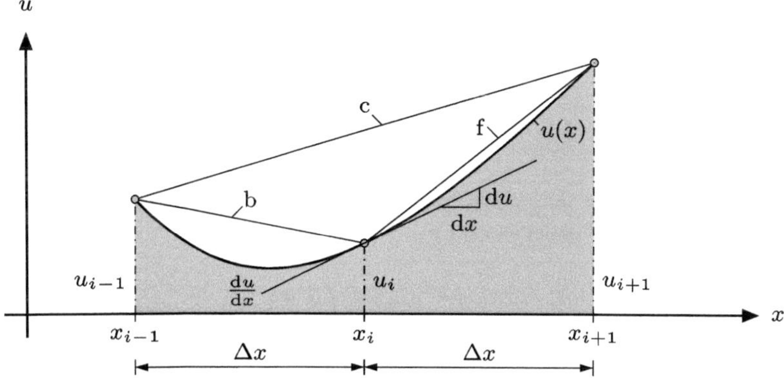

Fig. 2.2 Graphical representation of first order derivative approximations: **c** centered; **f** forward; **b** backward difference

$$u_{i+1} - u_i = \left(\frac{du}{dx}\right)_i \Delta x + \left(\frac{d^2u}{dx^2}\right)_i \frac{\Delta x^2}{2} + \cdots, \tag{2.8}$$

or

$$\left(\frac{du}{dx}\right)_i = \frac{u_{i+1} - u_i}{\Delta x} - \underbrace{\left(\frac{d^2u}{dx^2}\right)_i \Delta x - \cdots}_{O(\Delta x)}, \tag{2.9}$$

which gives an expression for the forward difference or forward Euler approximation of the first order derivative. In a similar way, Eq. (2.3) can be rearranged to obtain the backward difference or backward EULER approximation of the first order derivative as:

$$\left(\frac{du}{dx}\right)_i = \frac{u_i - u_{i-1}}{\Delta x} + \underbrace{\left(\frac{d^2u}{dx^2}\right)_i \Delta x - \cdots}_{O(\Delta x)}. \tag{2.10}$$

Graphical representations of the finite difference approximations of first order derivatives are shown in Fig. 2.2.

Further derivatives of $u(x)$ of different order can be derived if u is evaluated at nodes $i + 2$, $i - 2$ etc. through a Taylor's series about the node i. Finite difference formulae of the first order derivative with truncation error of order Δx^2, based on a forward or backward difference approximation, can be obtained in the following way: Let us consider in Eq. (2.3) the second order derivative, i.e.

$$u_{i-1} - u_i = -\left(\frac{du}{dx}\right)_i \Delta x + \frac{1}{2}\left(\frac{d^2u}{dx^2}\right)_i \Delta x^2 - \frac{1}{6}\left(\frac{d^3u}{dx^3}\right)_i \Delta x^3 + \cdots \tag{2.11}$$

or

$$\left(\frac{du}{dx}\right)_i = \frac{u_i - u_{i-1}}{\Delta x} + \frac{1}{2}\left(\frac{d^2 u}{dx^2}\right)_i \Delta x - \underbrace{\frac{1}{6}\left(\frac{d^3 u}{dx^3}\right)_i \Delta x^2 + \cdots,}_{O(\Delta x^2)} \tag{2.12}$$

where the second order derivative can be replaced by a backward approximation[3] with truncation error of order Δx^2, i.e.

$$\left(\frac{d^2 u}{dx^2}\right)_i = \frac{u_i - 2u_{i-1} + u_{i-2}}{\Delta x^2} + \left(\frac{d^3 u}{dx^3}\right)_i \Delta x - O(\Delta x^2). \tag{2.13}$$

Thus, the backward approximation of the first order derivative with truncation error of order Δx^2 is obtained as:

$$\left(\frac{du}{dx}\right)_i = \frac{u_i - u_{i-1}}{\Delta x} + \frac{u_i - 2u_{i-1} + u_{i-2}}{2\Delta x} + \frac{1}{2}\left(\left(\frac{d^3 u}{dx^3}\right)_i \Delta x^2 - O(\Delta x^3)\right)$$

$$= \frac{3u_i - 4u_{i-1} + u_{i-2}}{2\Delta x} + O(\Delta x^2). \tag{2.14}$$

Some common expressions for derivatives of different order and the respective truncation errors are summarized in Table 2.1.

2.2   Derivation Based on the Inner Product

It is also possible to derive the finite difference approximations based on the inner product, which reads for constant tensile stiffness (cf. Eq. (1.29)):

$$\int_0^L W(x)\left(EA\frac{d^2 u_x(x)}{dx^2} + p_x(x)\right) dx \stackrel{!}{=} 0. \tag{2.15}$$

For this alternative derivation, a special type of collocation method [2] can be applied. This so-called cell collocation method [4] considers a cell as shown in Fig. 2.3. Let us remind here that the point collocation method uses as weight function the Dirac delta function, i.e.,

$$\delta(x - x_i) = \begin{cases} 0 & \text{for } x \neq x_i \\ \infty & \text{for } x = x_i \end{cases}. \tag{2.16}$$

---

[3] This formulation can be obtained based on a Taylor's series expansion for $i - 2$ up to $\frac{d^6 u}{dx^6}$ and introducing a backward difference approximation of the first order derivative which contains the terms up to $\frac{d^6 u}{dx^6}$.

Table 2.1 Finite difference approximations for various differentiations, partly taken from [1, 3]

Derivative	Finite difference approximation	Type	Error
$\left(\dfrac{\mathrm{d}u}{\mathrm{d}x}\right)_i$	$\dfrac{u_{i+1} - u_i}{\Delta x}$	forward diff.	$O(\Delta x)$
	$\dfrac{-3u_i + 4u_{i+1} - u_{i+2}}{2\Delta x}$	"	$O(\Delta x^2)$
	$\dfrac{u_i - u_{i-1}}{\Delta x}$	backward diff.	$O(\Delta x)$
	$\dfrac{3u_i - 4u_{i-1} + u_{i-2}}{2\Delta x}$	"	$O(\Delta x^2)$
	$\dfrac{u_{i+1} - u_{i-1}}{2\Delta x}$	centered diff.	$O(\Delta x^2)$
$\left(\dfrac{\mathrm{d}^2u}{\mathrm{d}x^2}\right)_i$	$\dfrac{u_{i+2} - 2u_{i+1} + u_i}{\Delta x^2}$	forward diff.	$O(\Delta x)$
	$\dfrac{-u_{i+3} + 4u_{i+2} - 5u_{i+1} + 2u_i}{\Delta x^2}$	"	$O(\Delta x^2)$
	$\dfrac{u_i - 2u_{i-1} + u_{i-2}}{\Delta x^2}$	backward diff.	$O(\Delta x)$
	$\dfrac{2u_i - 5u_{i-1} + 4u_{i-2} - u_{i-3}}{\Delta x^2}$	"	$O(\Delta x^2)$
	$\dfrac{u_{i+1} - 2u_i + u_{i-1}}{\Delta x^2}$	centered diff.	$O(\Delta x^2)$
$\left(\dfrac{\mathrm{d}^3u}{\mathrm{d}x^3}\right)_i$	$\dfrac{u_{i+3} - 3u_{i+2} + 3u_{i+1} - u_i}{\Delta x^3}$	forward diff.	$O(\Delta x)$
	$\dfrac{-3u_{i+4} + 14u_{i+3} - 24u_{i+2} + 18u_{i+1} - 5u_i}{2\Delta x^3}$	"	$O(\Delta x^2)$
	$\dfrac{u_i - 3u_{i-1} + 3u_{i-2} - u_{i-3}}{\Delta x^3}$	backward diff.	$O(\Delta x)$
	$\dfrac{5u_i - 18u_{i-1} + 24u_{i-2} - 14u_{i-3} + 3u_{i-4}}{2\Delta x^3}$	"	$O(\Delta x^2)$
	$\dfrac{u_{i+2} - 2u_{i+1} + 2u_{i-1} - u_{i-2}}{2\Delta x^3}$	centered diff.	$O(\Delta x^2)$
$\left(\dfrac{\mathrm{d}^4u}{\mathrm{d}x^4}\right)_i$	$\dfrac{u_{i+4} - 4u_{i+3} + 6u_{i+2} - 4u_{i+1} + u_i}{\Delta x^4}$	forward diff.	$O(\Delta x)$
	$\dfrac{-2u_{i+5} + 11u_{i+4} - 24u_{i+3} + 26u_{i+2} - 14u_{i+1} + 3u_i}{\Delta x^4}$	"	$O(\Delta x^2)$
	$\dfrac{u_i - 4u_{i-1} + 6u_{i-2} - 4u_{i-3} + u_{i-4}}{\Delta x^4}$	backward diff.	$O(\Delta x)$
	$\dfrac{3u_i - 14u_{i-1} + 26u_{i-2} - 24u_{i-3} + 11u_{i-4} - 2u_{i-5}}{\Delta x^4}$	"	$O(\Delta x^2)$
	$\dfrac{u_{i+2} - 4u_{i+1} + 6u_i - 4u_{i-1} + u_{i-2}}{\Delta x^4}$	centered diff.	$O(\Delta x^2)$

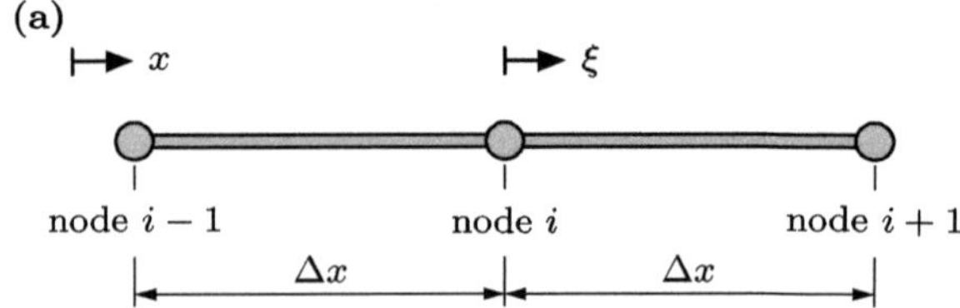

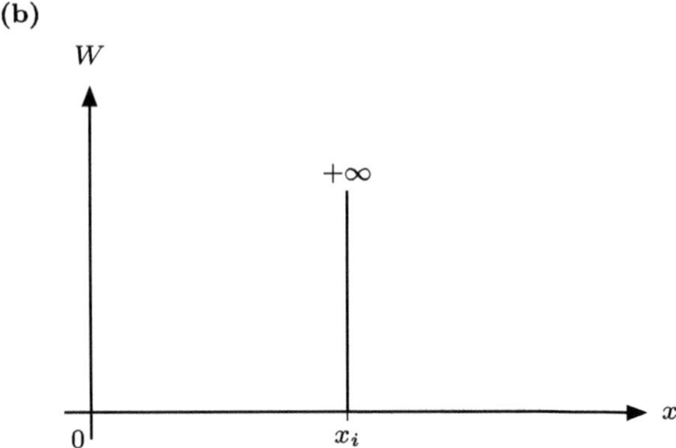

Fig. 2.3 a Typical cell for the cell collocation method and b appropriate weight function

To approximate, for example, a second order derivative $\frac{\mathrm{d}^2 u}{\mathrm{d}x^2}$, a local approximate function can be written as

$$u = \sum_{k=i-1}^{i} N_k \times u_k = N_{i-1}u_{i-1} + N_i u_i + N_{i+1}u_{i+1} , \qquad (2.17)$$

where $u_{i-1}, \ldots, u_{i+1}$ are the values of the function at the nodes. The interpolation functions N_k are given in natural coordinates $(-1 \leq \xi \leq 1)$ and can take the following quadratic form (cf. Fig. 2.4), [2]:

$$N_{i-1} = \frac{1}{2}\xi(\xi - 1) , \quad N_i = (1 - \xi)(1 + \xi) , \quad N_{i+1} = \frac{1}{2}\xi(1 + \xi) . \qquad (2.18)$$

The second-order derivative of the approximate function as given in Eq. (2.17) can be written under consideration of the coordinate transformation between the natural and Cartesian coordinate, i.e. $\xi = x/\Delta x$, or after an one-time differentiation, i.e $\mathrm{d}\xi/\mathrm{d}x = 1/\Delta x$ based on

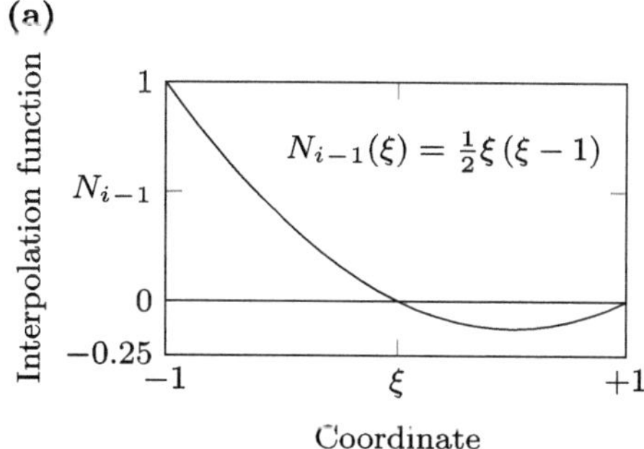

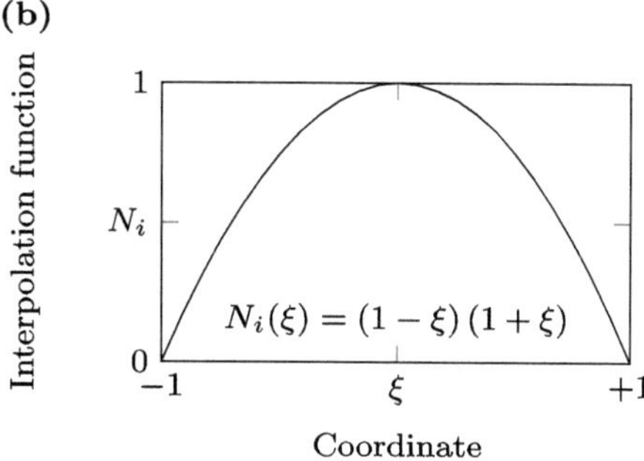

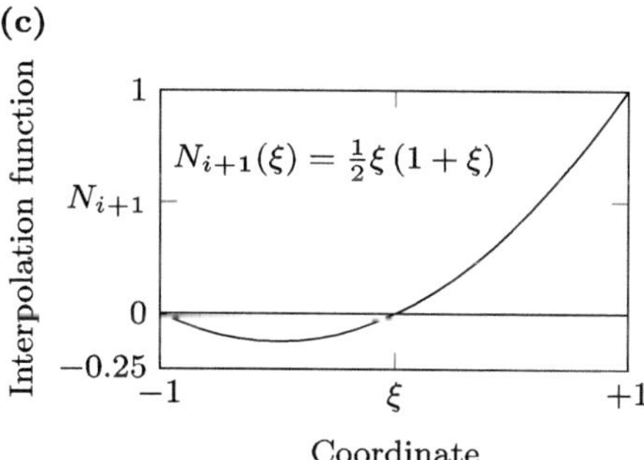

Fig. 2.4 Interpolation functions for cell collocation to derive a finite difference scheme: **a** $N_{i-1}(\xi)$, **b** $N_i(\xi)$, **c** $N_{i+1}(\xi)$, adapted from [2]

$$\frac{\mathrm{d}}{\mathrm{d}x} = \frac{\mathrm{d}}{\mathrm{d}\xi}\frac{\mathrm{d}\xi}{\mathrm{d}x} = \frac{1}{\Delta x}\frac{\mathrm{d}}{\mathrm{d}\xi}, \tag{2.19}$$

$$\frac{\mathrm{d}^2}{\mathrm{d}x^2} = \frac{\mathrm{d}^2}{\mathrm{d}\xi^2}\left(\frac{\mathrm{d}\xi}{\mathrm{d}x}\right)^2 = \frac{1}{\Delta x^2}\frac{\mathrm{d}^2}{\mathrm{d}\xi^2}, \tag{2.20}$$

as

$$\frac{\mathrm{d}^2 u}{\mathrm{d}x^2} = \frac{1}{\Delta x^2}\left(\frac{\mathrm{d}^2 N_{i-1}}{\mathrm{d}\xi^2}u_{i-1} + \frac{\mathrm{d}^2 N_i}{\mathrm{d}\xi^2}u_i + \frac{\mathrm{d}^2 N_{i+1}}{\mathrm{d}\xi^2}u_{i+1}\right)$$

$$= \frac{1}{\Delta x^2}(1u_{i-1} - 2u_i + 1u_{i+1}) \ . \tag{2.21}$$

Thus, the residual

$$r(x) = \frac{EA}{\Delta x^2}(1u_{i-1} - 2u_i + 1u_{i+1}) + p_i \neq 0 \tag{2.22}$$

can be used to formulate the collocation statement at node i as:

$$\int_{i-1}^{i+1} r W \mathrm{d}x = \int_{i-1}^{i+1}\left(\frac{EA}{\Delta x^2}(1u_{i-1} - 2u_i + 1u_{i+1}) + p_i\right)\delta(x - x_i)\mathrm{d}x \overset{!}{=} 0, \tag{2.23}$$

or after splitting the integral:

$$\int_{i-1}^{i+1}\frac{EA}{\Delta x^2}(1u_{i-1} - 2u_i + 1u_{i+1})\,\delta(x - x_i)\mathrm{d}x + \int_{i-1}^{i+1} p_i\delta(x - x_i)\mathrm{d}x \overset{!}{=} 0, \tag{2.24}$$

Under consideration of the properties of the Dirac delta function [5], i.e.,

$$\int_{-\infty}^{\infty}\delta(x - x_i)\,\mathrm{d}x = \int_{x_i-\varepsilon}^{x_i+\varepsilon}\delta(x - x_i)\,\mathrm{d}x = 1\,, \tag{2.25}$$

$$\int_{-\infty}^{\infty} f(x)\delta(x - x_i)\,\mathrm{d}x = \int_{x_i-\varepsilon}^{x_i+\varepsilon} f(x)\delta(x - x_i)\,\mathrm{d}x = f(x_i)\,, \tag{2.26}$$

the last equation gives the statement

$$\frac{EA}{\Delta x^2}(1u_{i-1} - 2u_i + 1u_{i+1}) = -p_i\,, \tag{2.27}$$

which is equivalent to the centered difference scheme as given in Table 2.1.

Fig. 2.5 a Typical subregion for the collocation by subregion method and **b** appropriate weight function, adapted from [2]

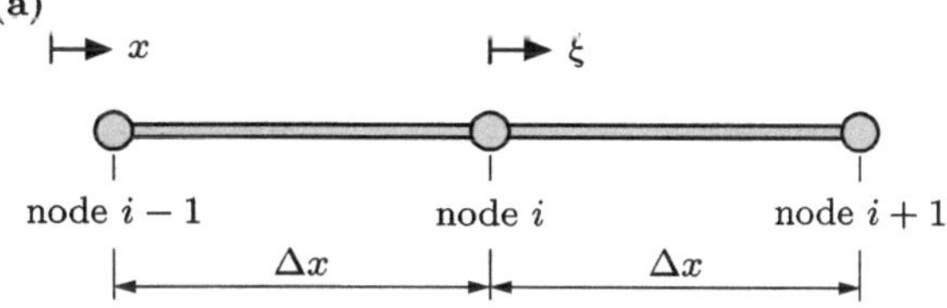

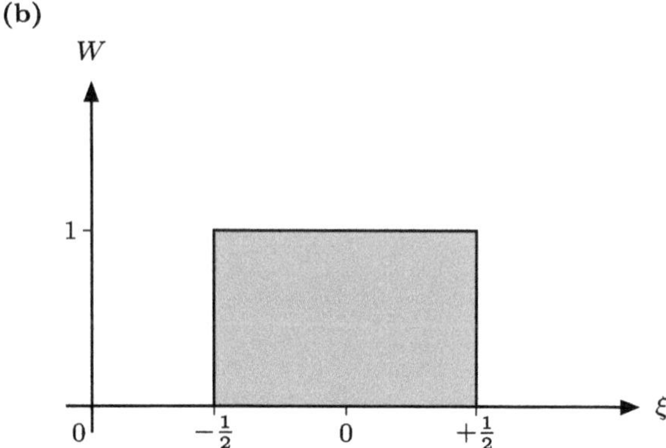

2.3 Derivation Based on the Weak Formulation

Instead of using the weighted residual method in the form of cell collocation, the method of collocation by subregions can be alternatively used to derive the finite difference method, cf. [2]. For this approach, the weight function is chosen as a step-type function as shown in Fig. 2.5.

Let us consider again the inner product, which reads for constant tensile stiffness (cf. Eq. (1.29)) as:

$$\int\limits_{-\frac{\Delta x}{2}}^{\frac{\Delta x}{2}} \left(EA \frac{\mathrm{d}^2 u}{\mathrm{d}x^2} + p_x(x) \right) W \,\mathrm{d}x = 0 \,. \tag{2.28}$$

Integrating by parts gives the following expression

$$\int\limits_{-\frac{\Delta x}{2}}^{\frac{\Delta x}{2}} EA \frac{\mathrm{d}^2 u}{\mathrm{d}x^2} W \,\mathrm{d}x = \left[EA \frac{\mathrm{d}u}{\mathrm{d}x} W \right]_{-\frac{\Delta x}{2}}^{\frac{\Delta x}{2}} - \int\limits_{-\frac{\Delta x}{2}}^{\frac{\Delta x}{2}} EA \frac{\mathrm{d}u}{\mathrm{d}x} \frac{\mathrm{d}W}{\mathrm{d}x} \,\mathrm{d}x + \int\limits_{-\frac{\Delta x}{2}}^{\frac{\Delta x}{2}} p_x(x) W \,\mathrm{d}x = 0 \,, \tag{2.29}$$

or rearranged as the weak formulation:

$$\int_{-\frac{\Delta x}{2}}^{\frac{\Delta x}{2}} EA \frac{du}{dx}\frac{dW}{dx} dx = \left[EA \frac{du}{dx} W \right]_{-\frac{\Delta x}{2}}^{\frac{\Delta x}{2}} + \int_{-\frac{\Delta x}{2}}^{\frac{\Delta x}{2}} p_x(x) W \, dx \,. \tag{2.30}$$

It holds for the step function that $W = 1$ and $\frac{dW}{dx} = 0$ and the last equation can be simplified to:

$$EA \left. \frac{du}{dx} \right|_{+\frac{\Delta x}{2}} - EA \left. \frac{du}{dx} \right|_{-\frac{\Delta x}{2}} = - \int_{-\frac{\Delta x}{2}}^{\frac{\Delta x}{2}} p_x(x) W \, dx \,. \tag{2.31}$$

To approximate the first order derivative $\frac{du}{dx}$, a local approximate function can be again written as in Eq. 2.17, i.e.

$$u = \sum_{k=i-1}^{i} N_k \times u_k = N_{i-1}u_{i-1} + N_i u_i + N_{i+1}u_{i+1} \,, \tag{2.32}$$

where $u_{i-1}, \ldots, u_{i+1}$ are the values of the function at the nodes. The interpolation functions N_k are given in natural coordinates $(-1 \leq \xi \leq 1)$ and can take again the quadratic form in Eq. (2.18) (cf. Fig. 2.4). The first-order derivative of the approximate function as given in Eq. (2.31) can be written under consideration of the coordinate transformation between the natural and Cartesian coordinate, i.e. $\xi = x/\Delta x$, or after an one-time differentiation, i.e $d\xi/dx = 1/\Delta x$ based on

$$\frac{d}{dx} = \frac{d}{d\xi}\frac{d\xi}{dx} = \frac{1}{\Delta x}\frac{d}{d\xi} \,, \tag{2.33}$$

as

$$\begin{aligned}
\frac{du}{dx} &= \frac{1}{\Delta x}\left(\frac{dN_{i-1}}{d\xi}u_{i-1} + \frac{dN_i}{d\xi}u_i + \frac{dN_{i+1}}{d\xi}u_{i+1} \right) \\
&= \frac{1}{\Delta x}\left(\frac{1}{2}(2\xi - 1)\,u_{i-1} + (-2\xi)\,u_i + \frac{1}{2}(1 + 2\xi)\,u_{i+1} \right) \,.
\end{aligned} \tag{2.34}$$

The last equation must be evaluated at the boundaries, i.e.

$$x = +\frac{\Delta x}{2} \rightarrow \xi = +\frac{1}{2} \,, \tag{2.35}$$

$$x = -\frac{\Delta x}{2} \rightarrow \xi = -\frac{1}{2} \,. \tag{2.36}$$

Considering in addition that the equivalent nodal force R_i, resulting from a distributed load, is in general given for an *inner*[4] node i as: $R_i = \int_{-\Delta x/2}^{\Delta x/2} p(\hat{x}) \mathrm{d}\hat{x}$, Eq. (2.31) can be written as

$$EA \left(\frac{u_{i+1} - u_i}{\Delta x} - \frac{u_i - u_{i-1}}{\Delta x} \right) = -R_i \,, \tag{2.37}$$

which corresponds to the centered difference scheme as given in Table 2.1.

References

1. K.-J. Bathe, *Finite Element Procedures* (Prentice-Hall, Upper Saddle River, 1996)
2. C.A. Brebbia, J.F.C. Telles, L.C. Wrobel, *Boundary Element Techniques: Theory and Applications* (Springer-Verlag, Berlin, 1984)
3. L. Collatz, *The Numerical Treatment of Differential Equations* (Springer-Verlag, Berlin, 1966)
4. P.C.M. Lau, C.A. Brebbia, The cell collocation method in continuum mechanics. Int. J. Mech. Sci. **20**, 83–95 (1978). https://doi.org/10.1016/0020-7403(78)90070-X
5. A. Öchsner, *(2014) Elasto-Plasticity of Frame Structure Elements: Modeling and Simulation of Rods and Beams* (Springer-Verlag, Berlin, 2014)
6. A. Öchsner, *Structural Mechanics with a Pen: A Guide to Solve Finite Difference Problems* (Springer, Cham, 2021)

---

[4] In the case of the boundary nodes ($i = 1$) or ($i = n$), the equivalent nodal loads must be calculated as $R_1 = \int_0^{\Delta x/2} p(\hat{x}) \mathrm{d}\hat{x}$ or $R_n = \int_{-\Delta x/2}^0 p(\hat{x}) \mathrm{d}\hat{x}$ in order to completely distribute the entire load $p(X)$ to the nodes.

Chapter 3
Finite Element Method

3.1 General Approach for One-Dimensional Bar Elements

Starting point for the derivation of the finite element formulation is the *weak formulation* as given in Eq. (1.32):

$$\int_0^L \frac{\mathrm{d}W(x)}{\mathrm{d}x}E(x)A(x)\frac{\mathrm{d}u_x(x)}{\mathrm{d}x}\mathrm{d}x = \left[W(x)E(x)A(x)\frac{\mathrm{d}u_x(x)}{\mathrm{d}x}\right]_0^L + \int_0^L W(x)p_x(x)\mathrm{d}x . \tag{3.1}$$

The symmetry in Eq. (3.1) with respect to the derivatives of the approximate solution $u_x(x)$ and the weight function $W(x)$ will guarantee in the following that a symmetric stiffness matrix is derived for the considered finite element. Let us assume for the following derivations that the tensile stiffness is constant, i.e. $E(x)A(x) \to EA$.

In order to continue the derivation of the principal finite element equation, the displacement $u_x(x)$ and the weight function $W(x)$ must be expressed by some functions. The common way to express the unknown function $u(x)$ in the scope of the finite element method is the so-called nodal approach. This approach states that the unknown displacement function within an element (superscript 'e') is given by

$$u^{\mathrm{e}}(x) = N^{\mathrm{T}}(x)\,u_{\mathrm{p}}^{\mathrm{e}} = \begin{bmatrix} N_1 & N_2 & \cdots & N_n \end{bmatrix} \times \begin{bmatrix} u_1 \\ u_2 \\ \vdots \\ u_n \end{bmatrix}, \tag{3.2}$$

where $u_{\mathrm{p}}^{\mathrm{e}}$ is the column matrix of n nodal unknown displacements and $N(x)$ is the column matrix of the *interpolation functions*. Thus, the displacement at any point inside an element is approximated based on nodal values and interpolation functions which distribute these displacements between the nodes in a certain way. Equation (3.2) illustrates a basic idea of the finite element method where the unknown

© The Author(s), under exclusive license to Springer Nature Switzerland AG 2026
A. Öchsner, *An Introduction to the Classical Approximation Methods in Applied Mechanics*, SpringerBriefs in Computational Mechanics,
https://doi.org/10.1007/978-3-032-06967-2_3

function is not approximated over the entire domain of the problem (in general Ω) but in a sub-domain (Ω^{e}), the so-called finite element. In a similar way as the unknown function, the weight function is approximated as

$$
W(x) = N(x)^{\mathrm{T}} \delta u_{\mathrm{p}} = \begin{bmatrix} N_1 & N_2 & \cdots & N_n \end{bmatrix} \times \begin{bmatrix} \delta u_1 \\ \delta u_2 \\ \vdots \\ \delta u_n \end{bmatrix}, \tag{3.3}
$$

where δu_i represents the so-called arbitrary or virtual displacements. It will be shown in the following that the virtual displacements occur on both sides of Eq. (3.1) and can be eliminated. Thus, these virtual displacements do not need a deeper consideration at this point of the derivation. Equation (3.1) requires the derivatives of $u_x(x)$ and $W(x)$ which can be written on the element level as:

$$
\frac{\mathrm{d}u^{\mathrm{e}}(x)}{\mathrm{d}x} = \frac{\mathrm{d}}{\mathrm{d}x} \left(N^{\mathrm{T}}(x)\, u_{\mathrm{p}} \right) = \frac{\mathrm{d}N^{\mathrm{T}}(x)}{\mathrm{d}x} u_{\mathrm{p}}, \tag{3.4}
$$

$$
\frac{\mathrm{d}W(x)}{\mathrm{d}x} = \frac{\mathrm{d}}{\mathrm{d}x} \left(N^{\mathrm{T}}(x)\, \delta u_{\mathrm{p}} \right) = \frac{\mathrm{d}N^{\mathrm{T}}(x)}{\mathrm{d}x} \delta u_{\mathrm{p}}. \tag{3.5}
$$

It should be noted here that the nodal unknowns and their virtual counterparts are constant values, i.e. not a function of x, and are therefore not affected by the differential operator. It is common in some references (e.g. [1, 4]) to introduce the matrix which contains the derivatives of the interpolation functions as a matrix denoted by $B = \frac{\mathrm{d}N(x)}{\mathrm{d}x}$. Thus, the derivatives can be be written as:

$$
\frac{\mathrm{d}u^{\mathrm{e}}(x)}{\mathrm{d}x} = B^{\mathrm{T}} u_{\mathrm{p}}, \tag{3.6}
$$

$$
\frac{\mathrm{d}W(x)}{\mathrm{d}x} = B^{\mathrm{T}} \delta u_{\mathrm{p}}. \tag{3.7}
$$

3.2   Linear Element Formulation

Let us consider in the following a bar element which is composed of two nodes as schematically shown in Fig. 3.1. Each node has only one degree of freedom, i.e., a displacement in the direction of the principal axis (cf. Fig. 3.1a) and each node can be only loaded by a single force acting in x-direction (cf. Fig. 3.1b).

Since there are only two nodes with two unknowns, the equation for the unknown displacement in the element and its virtual counterpart (cf. Eqs. (3.2) and (3.3)) are simplified to the following expressions:

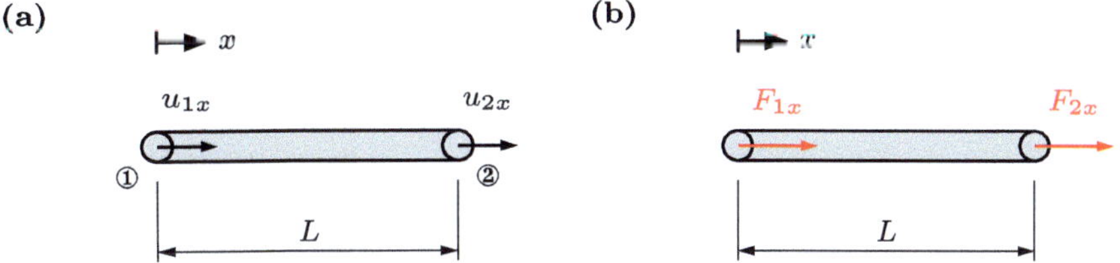

Fig. 3.1 Definition of the one-dimensional linear bar element: **a** deformations; **b** external loads. The nodes are symbolized by the two circles at the ends (○)

$$u^{e}(x) = \boldsymbol{N}^{\mathrm{T}}(x)\,\boldsymbol{u}_{\mathrm{p}} = \begin{bmatrix} N_1 & N_2 \end{bmatrix} \times \begin{bmatrix} u_1 \\ u_2 \end{bmatrix}, \tag{3.8}$$

and

$$W(x) = \boldsymbol{N}(x)^{\mathrm{T}}\delta\boldsymbol{u}_{\mathrm{p}} = \begin{bmatrix} N_1 & N_2 \end{bmatrix} \times \begin{bmatrix} \delta u_1 \\ \delta u_2 \end{bmatrix}, \tag{3.9}$$

or for the transposed of the weight function:

$$W^{\mathrm{T}}(x) = \left(\boldsymbol{N}(x)^{\mathrm{T}}\delta\boldsymbol{u}_{\mathrm{p}}\right)^{\mathrm{T}} = \delta\boldsymbol{u}_{\mathrm{p}}^{\mathrm{T}}\boldsymbol{N}(x), \tag{3.10}$$

$$\frac{\mathrm{d}W^{\mathrm{T}}(x)}{\mathrm{d}x} = \delta\boldsymbol{u}_{\mathrm{p}}^{\mathrm{T}}\frac{\mathrm{d}\boldsymbol{N}(x)}{\mathrm{d}x}. \tag{3.11}$$

Let us first consider in the following only the left-hand side of Eq. (3.1) in order to derive the expression for the elemental stiffness matrix $\boldsymbol{K}^{\mathrm{e}}$ of the linear bar element. Introduction of expressions (3.8) and (3.9) in the weak form gives

$$EA \int_{0}^{L} \left(\delta\boldsymbol{u}_{\mathrm{p}}^{\mathrm{T}}\frac{\mathrm{d}\boldsymbol{N}(x)}{\mathrm{d}x}\right)\left(\frac{\mathrm{d}\boldsymbol{N}^{\mathrm{T}}(x)}{\mathrm{d}x}\boldsymbol{u}_{\mathrm{p}}\right)\mathrm{d}x, \tag{3.12}$$

or under consideration that the column matrix of the nodal unknowns can be considered as constant as:

$$\delta\boldsymbol{u}_{\mathrm{p}}^{\mathrm{T}}\,EA\underbrace{\int_{0}^{L}\left(\frac{\mathrm{d}\boldsymbol{N}(x)}{\mathrm{d}x}\right)\left(\frac{\mathrm{d}\boldsymbol{N}^{\mathrm{T}}(x)}{\mathrm{d}x}\right)\mathrm{d}x}_{\boldsymbol{K}^{\mathrm{e}}}\boldsymbol{u}_{\mathrm{p}}. \tag{3.13}$$

It will be seen in the following that the expression $\delta\boldsymbol{u}_{\mathrm{p}}^{\mathrm{T}}$ can be 'canceled' with an identical expression on the right-hand side of Eq. (3.1) and $\boldsymbol{u}_{\mathrm{p}}$ represents the column matrix of the unknown nodal displacements. Under consideration of the $\boldsymbol{D}$-matrix, the stiffness matrix can be expressed in a more general way for constant tensile stiffness EA as:

$$\boldsymbol{K}^{\mathrm{e}} = EA \int_0^L \boldsymbol{B}\boldsymbol{B}^{\mathrm{T}} \mathrm{d}x .$$
(3.14)

In order to further evaluate Eq. (3.13), we can introduce the components of the derivatives to give:

$$EA \int_0^L \begin{bmatrix} \dfrac{\mathrm{d}N_1(x)}{\mathrm{d}x} \\ \dfrac{\mathrm{d}N_2(x)}{\mathrm{d}x} \end{bmatrix} \begin{bmatrix} \dfrac{\mathrm{d}N_1(x)}{\mathrm{d}x} & \dfrac{\mathrm{d}N_2(x)}{\mathrm{d}x} \end{bmatrix} \mathrm{d}x ,$$
(3.15)

or after the matrix multiplication as:

$$EA \int_0^L \begin{bmatrix} \dfrac{\mathrm{d}N_1(x)}{\mathrm{d}x}\dfrac{\mathrm{d}N_1(x)}{\mathrm{d}x} & \dfrac{\mathrm{d}N_1(x)}{\mathrm{d}x}\dfrac{\mathrm{d}N_2(x)}{\mathrm{d}x} \\ \dfrac{\mathrm{d}N_2(x)}{\mathrm{d}x}\dfrac{\mathrm{d}N_1(x)}{\mathrm{d}x} & \dfrac{\mathrm{d}N_2(x)}{\mathrm{d}x}\dfrac{\mathrm{d}N_2(x)}{\mathrm{d}x} \end{bmatrix} \mathrm{d}x .$$
(3.16)

Any further evaluation of this equation requires now that the functional expressions $N_1(x)$ and $N_2(x)$ are known. The simplest assumption that can be done is that the nodal values are linearly distributed within the element, from its value at the node to zero at the opposite node. For such a linear superposition, the interpolation functions can be assumed as shown in Fig. 3.2a, b.

The graphical interaction of interpolation functions with the nodal displacement values is shown in Fig. 3.3.

The derivatives of the interpolation functions can easily be calculated as

$$\frac{\mathrm{d}N_1(x)}{\mathrm{d}x} = -\frac{1}{L}, \quad \frac{\mathrm{d}N_2(x)}{\mathrm{d}x} = \frac{1}{L},$$
(3.17)

$$\frac{\mathrm{d}N_1(\xi)}{\mathrm{d}\xi} = -\frac{1}{2}, \quad \frac{\mathrm{d}N_2(\xi)}{\mathrm{d}\xi} = \frac{1}{2}.$$
(3.18)

Thus, the $\boldsymbol{B}$-matrix given in Eq. (3.6) takes the form:

$$\boldsymbol{B} = \frac{1}{L}\begin{bmatrix} -1 \\ 1 \end{bmatrix} .$$
(3.19)

The derivatives introduced into Eq. (3.20) give

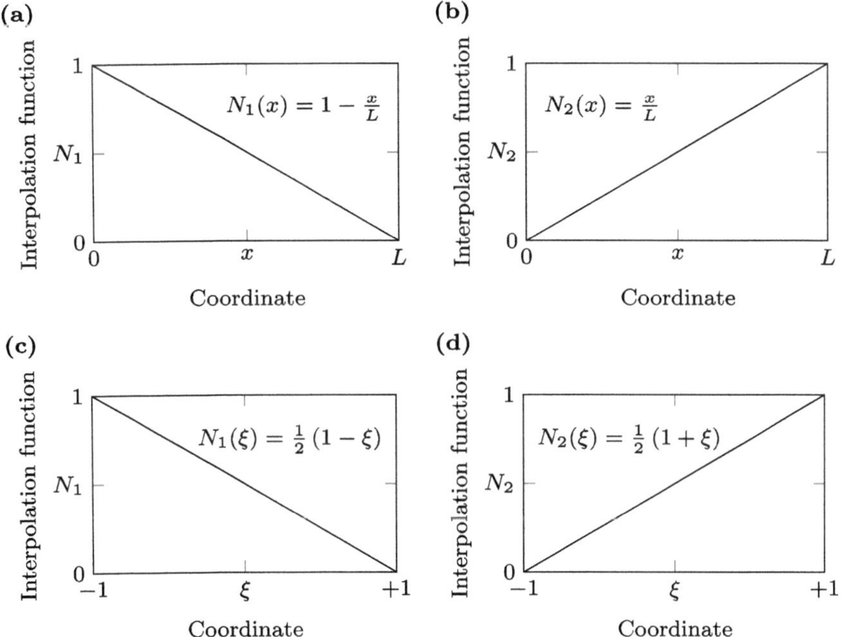

Fig. 3.2 Interpolation functions for the linear bar element: **a** and **b** physical coordinate (x); **c** and **d** natural coordinate (ξ)

$$EA \int_0^L \begin{bmatrix} \dfrac{1}{L^2} & -\dfrac{1}{L^2} \\ -\dfrac{1}{L^2} & \dfrac{1}{L^2} \end{bmatrix} \mathrm{d}x = \frac{EA}{L^2} \int_0^L \begin{bmatrix} 1 & -1 \\ -1 & 1 \end{bmatrix} \mathrm{d}x \,. \tag{3.20}$$

The integral in the last equation can be analytically integrated to obtain

$$\frac{EA}{L^2} \begin{bmatrix} x & -x \\ -x & x \end{bmatrix} \Bigg|_0^L = \frac{EA}{L^2} \begin{bmatrix} L & -L \\ -L & L \end{bmatrix} \tag{3.21}$$

and the stiffness matrix for a linear rod element is given by:

$$\boldsymbol{K}^{\mathrm{e}} = \frac{EA}{L} \begin{bmatrix} 1 & -1 \\ -1 & 1 \end{bmatrix}. \tag{3.22}$$

It must be noted here that an analytical integration as performed to obtain Eq. (3.21) cannot be performed in commercial finite element codes since they are written in traditional programming languages such as FORTRAN. Instead of the analytical integration, a numerical integration is performed (cf. [2]) where the integral is approxi-

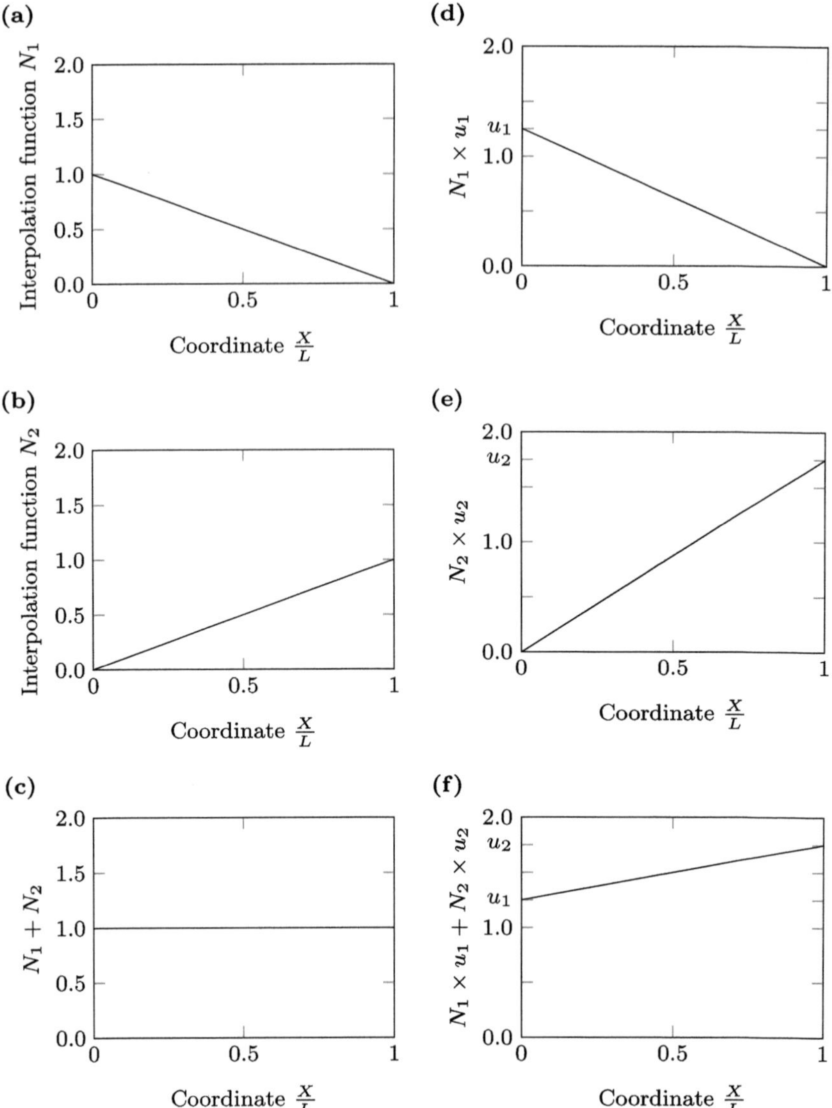

Fig. 3.3 Graphical interpretation of interpolation functions: **a–c** 'pure' interpolation functions; **d–e** weighted with nodal values

mated by the evaluation and weighting of functional values at so-called integration or Gauss points. To this end, the Cartesian coordinate x is transformed to the natural coordinate ξ ranging from -1 to 1. Depending on the origin of the Cartesian coordinate system, the transformation can be performed based on the relations given in Table 3.1. The integral in Eq. (3.20) can be written in terms of the natural coordinate ξ and approximated in terms of a Gauss-Legendre quadrature as:

$$\boldsymbol{K}^{\mathrm{e}} = \frac{EA}{L^2} \int_{-1}^{1} \begin{bmatrix} 1 & -1 \\ -1 & 1 \end{bmatrix} \frac{L}{2}\, \mathrm{d}\xi \approx \frac{EA}{L^2} \sum_{i=1}^{n} \begin{bmatrix} \cdots\cdots \\ \cdots\cdots \end{bmatrix} (\xi_i) w(\xi_i)\,, \tag{3.23}$$

where the matrix is to be evaluated at the n integration points and multiplied by certain weights w. Since the matrix is in this simple case only composed of constant values, it is sufficient to consider a one-point integration rule ($\xi = 0$, $w = 2$) to achieve the analytical result[1] as:

$$\boldsymbol{K}^{\mathrm{e}} = \frac{EA}{2L} \begin{bmatrix} 1 & -1 \\ -1 & 1 \end{bmatrix}\Bigg|_{\xi=0} \times \underbrace{2}_{w} = \frac{EA}{L} \begin{bmatrix} 1 & -1 \\ -1 & 1 \end{bmatrix}. \tag{3.24}$$

The transformation between Cartesian (x) and natural coordinates (ξ) as indicated in Table 3.1 can be further generalized. Let us assume for this purpose that the Cartesian coordinate can be interpolated in the following manner:

$$x(\xi) = \overline{N}_1(\xi) x_1 + \overline{N}_2(\xi) x_2\,, \tag{3.25}$$

where x_1 and x_2 are the coordinates of the start and end node in the elemental Cartesian coordinate system. The interpolation functions $\overline{N}_i(\xi)$ are—in the case of the coordinate approximation—called shape functions because they describe the geometry or shape of the element. Considering the shape functions in natural coordinates as given in Fig. 3.2 for the displacement interpolation (a so-called isoparametric formulation), the following expression for the derivative of the Cartesian coordinate with respect to the natural coordinate is obtained:

$$\frac{\mathrm{d}x(\xi)}{\mathrm{d}\xi} = \frac{\mathrm{d}\overline{N}_1(\xi)}{\mathrm{d}\xi} x_1 + \frac{\mathrm{d}\overline{N}_2(\xi)}{\mathrm{d}\xi} x_2 = -\frac{1}{2}x_1 + \frac{1}{2}x_2\,. \tag{3.26}$$

The last equation allows to reproduce the geometrical derivatives given in Table 3.1 or for any other location of the elemental Cartesian coordinate system. Equation (3.26) is also known as the general form of the Jacobian determinant and allows to perform the numerical integration of the stiffness matrix in natural coordinates as outlined in Eq. (A.16). The choice of the shape functions in Eq. (3.25) allows to distinguish

---

[1] It must be noted here that in the general case only an *approximation* of the integral can be obtained and that the exact, i.e. analytical solution, is reserved for simple cases.

Table 3.1 Transformation between Cartesian (x) and natural coordinates (ξ)

Configuration	Transformation
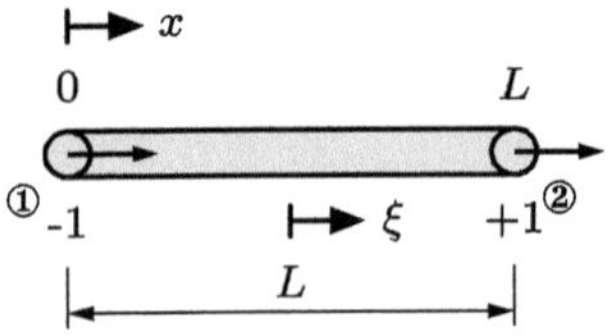 x 0 L ① -1 ξ $+1$ ② L	$\xi = \dfrac{2x}{L} - 1,$ $\dfrac{\mathrm{d}\xi}{\mathrm{d}x} = \dfrac{2}{L}$
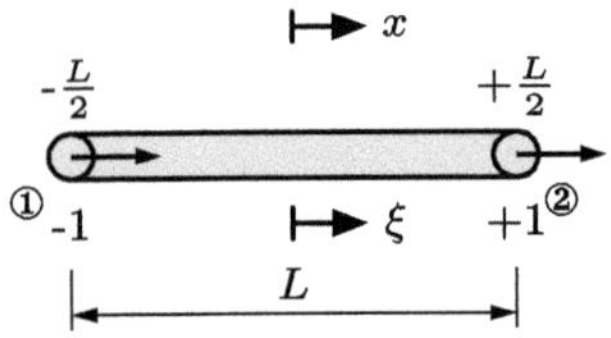 x $-\dfrac{L}{2}$ $+\dfrac{L}{2}$ ① -1 ξ $+1$ ② L	$\xi = \dfrac{2x}{L},$ $\dfrac{\mathrm{d}\xi}{\mathrm{d}x} = \dfrac{2}{L}$
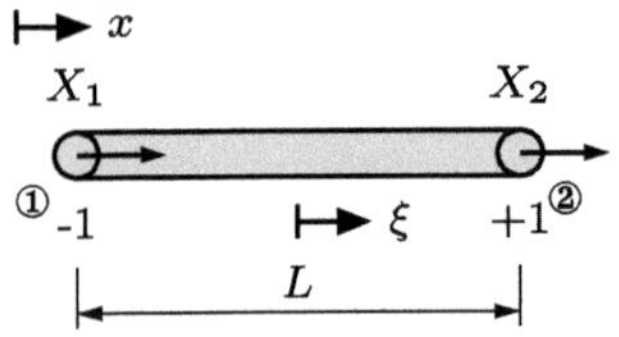 x X_1 X_2 ① -1 ξ $+1$ ② L	$\xi = \dfrac{2}{X_2 - X_1}(X - X_1) - 1,$ $\dfrac{\mathrm{d}\xi}{\mathrm{d}X} = \dfrac{2}{L}$

different element formulations. If the degree of the shape functions is equal to the degree of the interpolation functions, i.e. $\deg(\overline{N}) = \deg(N)$, a so-called isoparametric element formulation is obtained. If the degree of the shape functions is smaller than the degree of the interpolation functions, i.e. $\deg(\overline{N}) < \deg(N)$, a so-called subparametric element formulation is obtained. A larger degree of the shape functions compared to the interpolation functions, i.e. $\deg(\overline{N}) > \deg(N)$, gives a so-called superparametric element formulation.

Let us summarize here in a systematic manner the major steps which are required to calculate the elemental stiffness matrix of a linear bar element.

❶ Introduce an elemental coordinate system (x).

❷ Express the coordinates (x_i) of the corner nodes i $(i = 1, 2)$ in this elemental coordinate system.

❸ Calculate the partial derivative of the Cartesian (x) coordinate with respect to the natural (ξ) coordinate, see Eq. (3.26):

$$\frac{\mathrm{d}x(\xi)}{\mathrm{d}\xi} = J = -\frac{1}{2}x_1 + \frac{1}{2}x_2 \,.$$

❹ Calculate the partial derivative of the natural (ξ) coordinate with respect to the Cartesian (x) coordinate, see Eq. (A.25):

$$\frac{\mathrm{d}\xi}{\mathrm{d}x} = \frac{1}{J} \,.$$

❺ Calculate the $\boldsymbol{B}$-matrix and its transposed, see Eqs. (3.14)–(3.15):

$$\boldsymbol{B}^{\mathrm{T}} = \left[\frac{\mathrm{d}N_1(x)}{\mathrm{d}x} \ \ \frac{\mathrm{d}N_2(x)}{\mathrm{d}x}\right] \,,$$

where the partial derivatives are $\frac{\mathrm{d}N_1(x)}{\mathrm{d}x} = \frac{\mathrm{d}N_1(\xi)}{\mathrm{d}\xi}\frac{\mathrm{d}\xi}{\mathrm{d}x}, \dots$ and the derivatives of the interpolation functions are given in Eq. (3.18), i.e., $\frac{\partial N_1(\xi)}{\partial \xi} = -\frac{1}{2}, \dots$

❻ Calculate the triple matrix product $\boldsymbol{B}\boldsymbol{C}^{\mathrm{T}}\boldsymbol{B}$, where the elasticity matrix $\boldsymbol{C}$ is given in this special case as the scalar YOUNG's modulus E.

❼ Perform the numerical integration based on a 1-point integration rule:

$$\int_V (\boldsymbol{B}\boldsymbol{C}\boldsymbol{B}^{\mathrm{T}})\mathrm{d}V = \boldsymbol{B}E\boldsymbol{B}^{\mathrm{T}}J \times 2 \times A\Big|_{(0)} \,.$$

❽ $\boldsymbol{K}$ obtained.

Let us now consider the right-hand side of Eq. (3.1) in order to derive the expression for the elemental load column matrix $\boldsymbol{f}^{\mathrm{e}}$ of the linear bar element. The first part of the right-hand side, i.e.

$$EA\left[W^{\mathrm{T}}(x)\frac{\mathrm{d}u(x)}{\mathrm{d}x}\right]_0^L \tag{3.27}$$

results with the definition of the weight function according to Eq. (3.9) in

$$EA\left[\delta u_{\mathrm{p}}^{\mathrm{T}}N(x)\frac{\mathrm{d}u(x)}{\mathrm{d}x}\right]_0^L, \tag{3.28}$$

$$\delta u_{\mathrm{p}}^{\mathrm{T}}EA\left[\begin{bmatrix}N_1\\N_2\end{bmatrix}\frac{\mathrm{d}u(x)}{\mathrm{d}x}\right]_0^L. \tag{3.29}$$

The virtual displacements $\delta u_{\mathrm{p}}^{\mathrm{T}}$ in the last equation can be 'canceled' with a corresponding expression in Eq. (3.13). Furthermore, the last equation constitutes a system of two equations which must be evaluated at the integration boundaries, i.e. at $x = 0$ and $x = L$. The first equation reads:

$$\left(N_1EA\frac{\mathrm{d}u}{\mathrm{d}x}\right)_{x=L} - \left(N_1EA\frac{\mathrm{d}u}{\mathrm{d}x}\right)_{x=0}. \tag{3.30}$$

This gives under consideration of the boundary values of the interpolation functions, i.e. $N_1(L) = 0$ and $N_1(0) = 1$, the following statement:

$$-EA\frac{\mathrm{d}u}{\mathrm{d}x}\bigg|_{x=0} \overset{(1.16)}{=} -N_x(x = 0). \tag{3.31}$$

A corresponding expression can be derived for the second equation as:

$$EA\frac{\mathrm{d}u}{\mathrm{d}x}\bigg|_{x=L} \overset{(1.16)}{=} N_x(x = L). \tag{3.32}$$

It must be noted here that the forces N_x are the internal reactions according to Fig. 1.5. The external loads with their positive directions according to Fig. 3.1b can be obtained from the internal loads by inverting the sign at the left-hand boundary and by maintaining the positive direction of the internal reaction at the right-hand boundary. This can easily be shown by balancing the internal and external forces at each boundary node. Thus, the contribution to the load matrix due to single *external* forces F_i at the nodes is expressed by:

$$f_F^{\mathrm{e}} = \begin{bmatrix}F_{1x}\\F_{2x}\end{bmatrix}. \tag{3.33}$$

The second part of Eq. (1.32), i.e. after 'canceling' of the virtual displacements $\delta \boldsymbol{u}^{\mathrm{T}}$

$$\int_0^L \boldsymbol{N}(x)\, p(x)\, \mathrm{d}x \tag{3.34}$$

represents the general rule to determine equivalent nodal loads in the case of arbitrarily distributed loads $p(x)$. As an example, the evaluation of Eq. (3.34) for a constant load p results in the following load matrix:

$$\boldsymbol{f}_p^{\mathrm{e}} = p \int_0^L \begin{bmatrix} N_1 \\ N_2 \end{bmatrix} \mathrm{d}x = \frac{pL}{2} \begin{bmatrix} 1 \\ 1 \end{bmatrix}. \tag{3.35}$$

Based on the derived results, the principal finite element equation for a single linear bar element with constant axial tensile stiffness EA can be expressed in a general form as

$$\boldsymbol{K}^{\mathrm{e}} \boldsymbol{u}_{\mathrm{p}}^{\mathrm{e}} = \boldsymbol{f}^{\mathrm{e}}, \tag{3.36}$$

or in components as:

$$\frac{EA}{L} \begin{bmatrix} 1 & -1 \\ -1 & 1 \end{bmatrix} \begin{bmatrix} u_{1x} \\ u_{2x} \end{bmatrix} = \begin{bmatrix} F_{1x} \\ F_{2x} \end{bmatrix} + \int_0^L \begin{bmatrix} N_1 \\ N_2 \end{bmatrix} p_x(x)\, \mathrm{d}x. \tag{3.37}$$

3.3 Quadratic Element Formulation

Let us consider now a bar element which is composed of three nodes as schematically shown in Fig. 3.4. Each node has again only one degree of freedom, i.e. a displacement in x-direction and each node can be only loaded by a single force acting along the x-axis. It is assumed in the following that the second node is exactly located in the middle, i.e. at $x = \frac{L}{2}$, of the element.

Since there are now three nodes with three unknowns, the equation for the unknown displacement in the element and its virtual counterpart (cf. Eqs. (3.2) and (3.3)) are now given by the expressions:

$$u^{\mathrm{e}}(x) = \boldsymbol{N}^{\mathrm{T}}(x)\, \boldsymbol{u}_{\mathrm{p}} = \begin{bmatrix} N_1 & N_2 & N_3 \end{bmatrix} \times \begin{bmatrix} u_1 \\ u_2 \\ u_3 \end{bmatrix}, \tag{3.38}$$

and

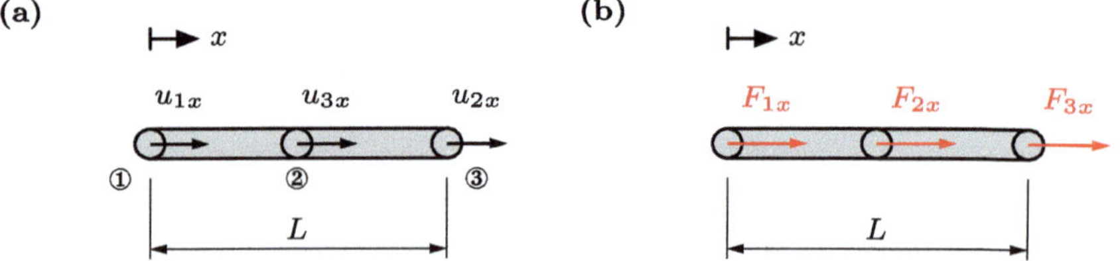

Fig. 3.4 Definition of the one-dimensional quadratic rod element: **a** deformations; **b** external loads. The nodes are symbolized by circles at the ends and in the middle ($\bigcirc$)

$$W(x) = N(x)^{\mathrm{T}}\delta u_{\mathrm{p}} = \begin{bmatrix} N_1 & N_2 & N_3 \end{bmatrix} \times \begin{bmatrix} \delta u_1 \\ \delta u_2 \\ \delta u_3 \end{bmatrix}. \tag{3.39}$$

Similar as in Eq. (3.20), the elemental stiffness matrix can be expressed before evaluating the integral as:

$$\boldsymbol{K}^{\mathrm{e}} = EA \int_0^L \begin{bmatrix} \dfrac{\mathrm{d}N_1(x)}{\mathrm{d}x}\dfrac{\mathrm{d}N_1(x)}{\mathrm{d}x} & \dfrac{\mathrm{d}N_1(x)}{\mathrm{d}x}\dfrac{\mathrm{d}N_2(x)}{\mathrm{d}x} & \dfrac{\mathrm{d}N_1(x)}{\mathrm{d}x}\dfrac{\mathrm{d}N_3(x)}{\mathrm{d}x} \\[2ex] \dfrac{\mathrm{d}N_2(x)}{\mathrm{d}x}\dfrac{\mathrm{d}N_1(x)}{\mathrm{d}x} & \dfrac{\mathrm{d}N_2(x)}{\mathrm{d}x}\dfrac{\mathrm{d}N_2(x)}{\mathrm{d}x} & \dfrac{\mathrm{d}N_2(x)}{\mathrm{d}x}\dfrac{\mathrm{d}N_3(x)}{\mathrm{d}x} \\[2ex] \dfrac{\mathrm{d}N_3(x)}{\mathrm{d}x}\dfrac{\mathrm{d}N_1(x)}{\mathrm{d}x} & \dfrac{\mathrm{d}N_3(x)}{\mathrm{d}x}\dfrac{\mathrm{d}N_2(x)}{\mathrm{d}x} & \dfrac{\mathrm{d}N_3(x)}{\mathrm{d}x}\dfrac{\mathrm{d}N_3(x)}{\mathrm{d}x} \end{bmatrix} \mathrm{d}x. \tag{3.40}$$

The interpolation functions N_i in this case[2] are given by quadratic equations as shown in Fig. 3.5 in physical and natural coordinates.

From the functional expressions given in Fig. 3.5, the derivatives are obtained as $\frac{\mathrm{d}N_1}{\mathrm{d}x} = -\frac{3}{L} + \frac{4x}{L^2}$, $\frac{\mathrm{d}N_2}{\mathrm{d}x} = \frac{4}{L} - \frac{8x}{L^2}$, and $\frac{\mathrm{d}N_3}{\mathrm{d}x} = -\frac{1}{L} + \frac{4x}{L^2}$ and Eq. (3.40) can be evaluated by analytical or numerical integration to give the elemental stiffness matrix of the quadratic rod element as:

$$\boldsymbol{K}^{\mathrm{e}} = \frac{EA}{3L} \begin{bmatrix} 7 & -8 & 1 \\ -8 & 16 & -8 \\ 1 & -8 & 7 \end{bmatrix}. \tag{3.41}$$

Furthermore, it should be noted that the $\boldsymbol{B}$-matrix, cf. Eq. (3.6), takes the following form for the quadratic rod element:

---

[2] A formal derivation of the functional expressions is presented in [3].

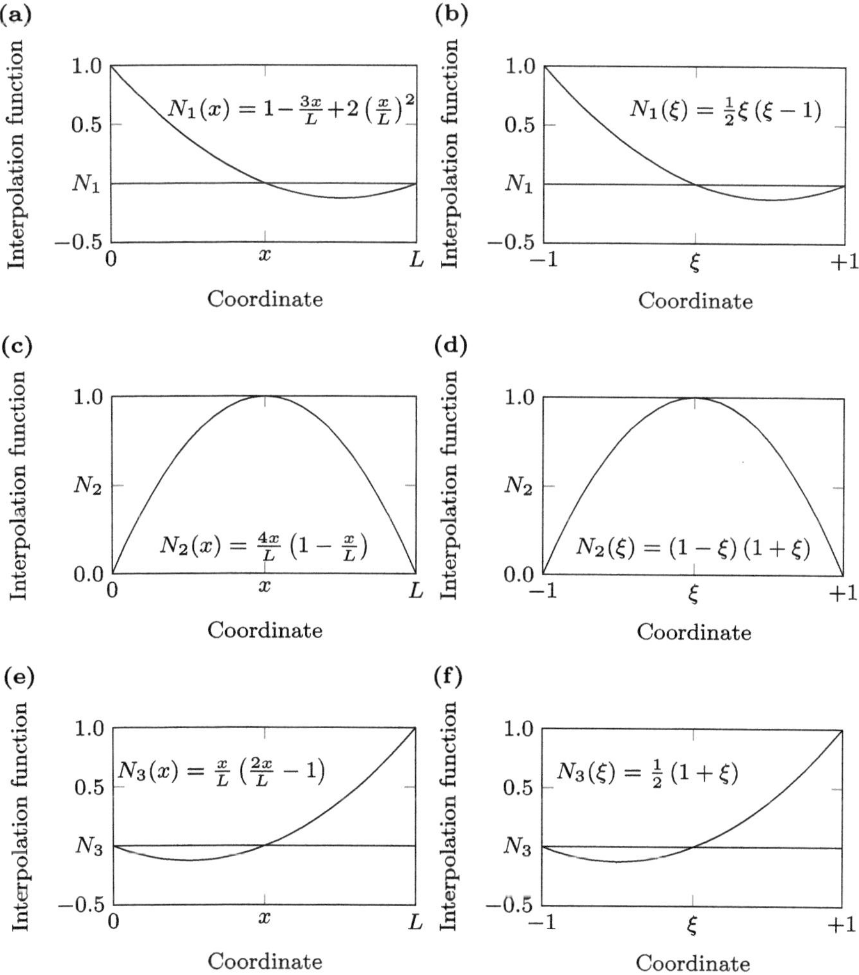

Fig. 3.5 Interpolation functions for the quadratic rod element with equidistant nodes: **a, c, e** physical coordinate (x); **b, d, f** natural coordinate (ξ)

$$\boldsymbol{B} = \frac{1}{L}\begin{bmatrix} -3 + \frac{4x}{L} \\ 4 - \frac{8x}{L} \\ -1 + \frac{4x}{L} \end{bmatrix} = \frac{1}{L}\begin{bmatrix} -1 + 2\xi \\ -4\xi \\ 1 + 2\xi \end{bmatrix}. \tag{3.42}$$

Let us summarize here in a systematic manner the major steps which are required to calculate the elemental stiffness matrix of a quadratic bar element.

❶ Introduce an elemental coordinate system (x).

❷ Express the coordinates (x_i) of the corner nodes i $(i = 1, 2, 3)$ in this elemental coordinate system.

❸ Calculate the partial derivative of the Cartesian (x) coordinate with respect to the natural (ξ) coordinate, (let us assume here an isoparametric element formulation with $\overline{N}_i = N_i$):

$$\frac{\mathrm{d}x(\xi)}{\mathrm{d}\xi} = J = \left(\xi - \frac{1}{2}\right) x_1 + (-2\xi)\, x_2 + \left(\frac{1}{2} + \xi\right) x_3 \,.$$

❹ Calculate the partial derivative of the natural (ξ) coordinate with respect to the Cartesian (x) coordinate:

$$\frac{\mathrm{d}\xi}{\mathrm{d}x} = \frac{1}{J} \,.$$

❺ Calculate the $\boldsymbol{B}$-matrix and its transposed:

$$\boldsymbol{B}^{\mathrm{T}} = \left[\frac{\mathrm{d}N_1(\xi)}{\mathrm{d}\xi}\frac{\mathrm{d}\xi}{\mathrm{d}x} \quad \frac{\mathrm{d}N_2(\xi)}{\mathrm{d}\xi}\frac{\mathrm{d}\xi}{\mathrm{d}x} \quad \frac{\mathrm{d}N_3(\xi)}{\mathrm{d}\xi}\frac{\mathrm{d}\xi}{\mathrm{d}x}\right] \,.$$

❻ Calculate the triple matrix product $\boldsymbol{BCB}^{\mathrm{T}}$, where the elasticity matrix $\boldsymbol{C}$ is given in this special case as the scalar Young's modulus E.

❼ Perform the numerical integration based on a 2-point integration rule:

$$\int_V (\boldsymbol{BCB}^{\mathrm{T}})\mathrm{d}V = \boldsymbol{B}E\boldsymbol{B}^{\mathrm{T}}J \times 1 \times A\Big|_{(\xi_1=-1/\sqrt{3})}$$

$$+ \boldsymbol{B}E\boldsymbol{B}^{\mathrm{T}}J \times 1 \times A\Big|_{(\xi_2=+1/\sqrt{3})} \,.$$

❽ $\boldsymbol{K}$ obtained.

The right-hand side of Eq. (3.1) can be treated in a similar way as in Sect. 3.2 to obtain the elemental load matrix in the form of:

$$\boldsymbol{f}^{\mathrm{e}} = \boldsymbol{f}^{\mathrm{e}}_F + \boldsymbol{f}^{\mathrm{e}}_p = \begin{bmatrix} F_{1x} \\ F_{2x} \\ F_{3x} \end{bmatrix} + \int_0^L \begin{bmatrix} N_1 \\ N_2 \\ N_3 \end{bmatrix} p(x)\,\mathrm{d}x \,, \tag{3.43}$$

where the expression with the integral represents the general rule to determine equivalent nodal loads in the case of arbitrarily distributed loads $p(x)$. As an example, the evaluation of this integral for a constant load $p_0 = \text{const.}$ results in the following load matrix:

$$\boldsymbol{f}_p^{\mathrm{e}} = p_0 \int_0^L \begin{bmatrix} N_1 \\ N_2 \\ N_3 \end{bmatrix} \mathrm{d}x = \frac{p_0 L}{6} \begin{bmatrix} 1 \\ 4 \\ 1 \end{bmatrix}. \tag{3.44}$$

Based on the derived results, the principal finite element equation for a single quadratic rod element with constant axial tensile stiffness EA can be expressed in components as:

$$\frac{EA}{3L} \begin{bmatrix} 7 & -8 & 1 \\ -8 & 16 & -8 \\ 1 & -8 & 7 \end{bmatrix} \begin{bmatrix} u_{1x} \\ u_{2x} \\ u_{3x} \end{bmatrix} = \begin{bmatrix} F_{1x} \\ F_{2x} \\ F_{3x} \end{bmatrix} + \int_0^L \begin{bmatrix} N_1 \\ N_2 \\ N_3 \end{bmatrix} p(x)\, \mathrm{d}x. \tag{3.45}$$

References

1. R.D. Cook, D.S. Malkus, M.E. Plesha, R.J. Witt, *Concepts and Applications of Finite Element Analysis* (John Wiley & Sons, New York, 2002)
2. R.H. MacNeal, *Finite Elements: Their Design and Performance* (Marcel Dekker, New York, 1994)
3. A. Öchsner, *Computational Statics and Dynamics: An Introduction Based on the Finite Element Method* (Springer, Cham, 2023)
4. O.C. Zienkiewicz, R.L. Taylor, *The Finite Element Method. Vol. 1: The Basis* (Butterworth-Heinemann, Oxford, 2000)

Chapter 4
Finite Volume Method

To derive the finite volume method, one has to divide the entire volume Ω of a structure into discrete control volumes Ω_i [1, 2]. Let us consider a number of nodes in the one-dimensional domain space, see Fig. 4.1a. The boundaries (or faces) of control volumes are positioned mid-way (i.e., at locations $i - \frac{1}{2}$ and $i + \frac{1}{2}$) between neighboring nodes. Thus each node is surrounded by a finite volume or cell Ω_i.

Let us consider the differential equation of a one-dimensional bar in a slightly different manner than provided in Eq. (1.19), i.e.,

$$E\frac{\mathrm{d}^2 u}{\mathrm{d}x^2} + b_x(x) = 0 , \tag{4.1}$$

where $b_x(x)$ is the distributed load per unit volume $(b_x(x) = p_x(x)/A)$. Now we consider the inner product, i.e., integrating the differential equation over the finite volume at its nodal point i, which reads (cf. Eq. (1.29)) as:

$$\int_{\Omega_i} W \left(E\frac{\mathrm{d}^2 u}{\mathrm{d}x^2} + b_x(x) \right) \, \mathrm{d}\Omega = 0 , \tag{4.2}$$

or for constant cross sectional areas, meaning $\mathrm{d}\Omega = A\mathrm{d}x$:

$$\int_{-\frac{\Delta x}{2}}^{\frac{\Delta x}{2}} W \left(EA\frac{\mathrm{d}^2 u}{\mathrm{d}x^2} + p_x(x) \right) \, \mathrm{d}x = 0 . \tag{4.3}$$

© The Author(s), under exclusive license to Springer Nature Switzerland AG 2026
A. Öchsner, *An Introduction to the Classical Approximation Methods in Applied Mechanics*, SpringerBriefs in Computational Mechanics,
https://doi.org/10.1007/978-3-032-06967-2_4

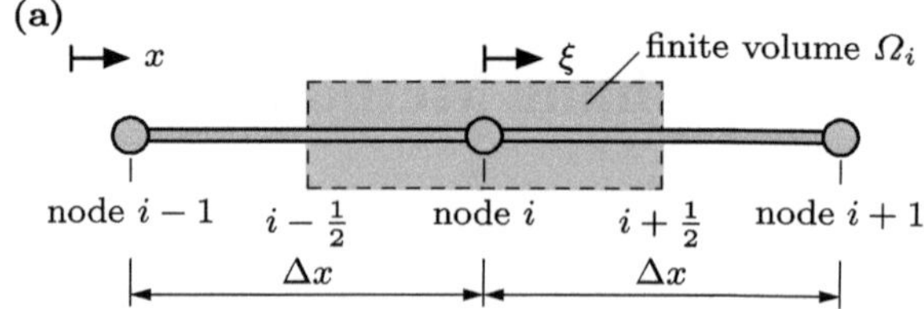

Fig. 4.1 a Typical finite volume Ω_i for the collocation by subregion method and **b** appropriate weight function

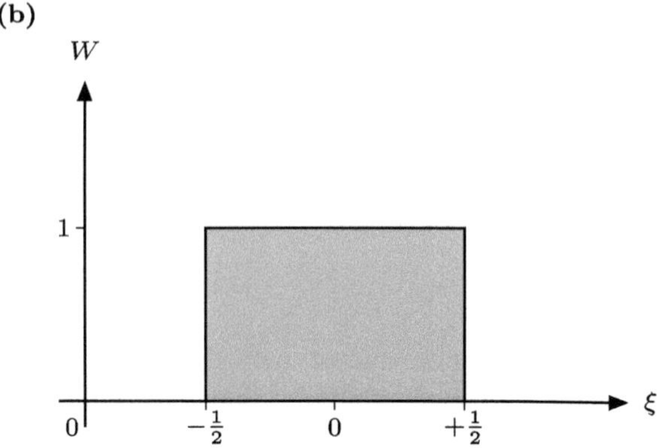

The following derivation is now identical to the approach presented for the finite difference method in Sect. 2.3. Thus, the weak formulation is obtained as

$$\int_{-\frac{\Delta x}{2}}^{\frac{\Delta x}{2}} EA \frac{du}{dx}\frac{dW}{dx}\, dx = \left[EA \frac{du}{dx} W \right]_{-\frac{\Delta x}{2}}^{\frac{\Delta x}{2}} + \int_{-\frac{\Delta x}{2}}^{\frac{\Delta x}{2}} p_x(x) W\, dx\,, \qquad (4.4)$$

and finally again the centered difference scheme as follows:

$$EA \left(\frac{u_{i+1} - u_i}{\Delta x} - \frac{u_i - u_{i-1}}{\Delta x} \right) = -R_i\,, \qquad (4.5)$$

where the equivalent nodal force R_i, resulting from a distributed load, is in general given for an *inner*[1] node i as: $R_i = \int_{-\Delta x/2}^{\Delta x/2} p(\hat{x})d\hat{x}$.

---

[1] In the case of the boundary nodes $(i = 1)$ or $(i = n)$, the equivalent nodal loads must be calculated as $R_1 = \int_0^{\Delta x/2} p(\hat{x})d\hat{x}$ or $R_n = \int_{-\Delta x/2}^{0} p(\hat{x})d\hat{x}$ in order to completely distribute the entire load $p(X)$ to the nodes.

References

1. R.J. LeVeque, *Finite Difference Methods for Ordinary and Partial Differential Equations: Steady-State and Time-Dependent Problems* (Society for Industrial and Applied Mathematics SIAM, Philadelphia, 2007)
2. R. Petrova, *Finite Volume Method: Powerful Means of Engineering Design* (InTech, Rijeka, 2012)

Chapter 5
Boundary Element Method

Starting point for the derivation of the boundary element formulation is the *inverse formulation* as given in Eq. (1.35):

$$\int_0^L \frac{\mathrm{d}^2 W(x)}{\mathrm{d}x^2} E(x)A(x)u_x(x)\mathrm{d}x = \left[\frac{\mathrm{d}W(x)}{\mathrm{d}x}E(x)A(x)u_x(x)\right]_0^L$$
$$- \left[W(x)E(x)A(x)\frac{\mathrm{d}u_x(x)}{\mathrm{d}x}\right]_0^L \tag{5.1}$$
$$- \int_0^L W(x)p_x(x)\mathrm{d}x\,.$$

The characteristic of the inverse formulation is that the differential operator is completely shifted to the weight function $W(x)$. The next step is to choose the weight function $W(x)$. In the scope of the boundary element method, the weight function is chosen to be a fundamental solution: $W(x) = W^*(x)$ [1, 3]. This fundamental or Kelvin solution $W^*(x)$ is in the considered case the deformation which is observed at some point x of an infinite bar due to a unit force at some distant point ξ in the bar. Alternatively, it is that function which satisfies the differential equation with right-hand side zero at every point of an infinite bar except the force point at which the right-hand side is infinite. Applied to our problem, the fundamental solution is defined by

$$\frac{\mathrm{d}^2 W^*(x)}{\mathrm{d}x^2} = \delta(x - \xi)\,, \tag{5.2}$$

A. Öchsner, *An Introduction to the Classical Approximation Methods in Applied Mechanics*, SpringerBriefs in Computational Mechanics, https://doi.org/10.1007/978-3-032-06967-2_5

where the Dirac delta function $\delta(x, \xi)$ is defined in a simple manner as:

$$\delta(x - \xi) = \begin{cases} \infty & \text{for } x = \xi \\ 0 & \text{for } x \neq \xi \end{cases} . \tag{5.3}$$

In the scope of the boundary element method, the selection or sifting property, i.e.

$$\int_{x_1}^{x_2} f(x)\delta(x - \xi)\mathrm{d}x = f(\xi), \tag{5.4}$$

is the main property which is used for the theoretical derivations. Equation (5.2) can be multiplied with the field variable u and integrated to obtain, under consideration of the selection property, the following statement:

$$\int_0^L \frac{\mathrm{d}^2 W^*(x)}{\mathrm{d}X^2} u(x)\mathrm{d}x = \int_0^L \delta(x, \xi)u(x)\mathrm{d}x = u(\xi), \tag{5.5}$$

which leads with Eq. (5.1) to the so-called representation formula

$$u(\xi) = -\int_0^L \frac{p(x)}{EA} W^* \mathrm{d}x - \left[\frac{\mathrm{d}u}{\mathrm{d}x} W^*\right]_0^L + \left[u \frac{\mathrm{d}W^*}{\mathrm{d}x}\right]_0^L . \tag{5.6}$$

Last equation yields the value $u(\xi)$ inside the domain, i.e. the bar, if the boundary solution is known.

In order to continue the derivation, the fundamental solution must be known. For the condition given in Eq. (5.2), the fundamental solution is given by:

$$W^*(x) = \frac{1}{2}|x - \xi| . \tag{5.7}$$

However, it must be noted here that no fundamental solution can be obtained for some differential operators. The graphical representation of Eq. (5.7) is shown in Fig. 5.1 where it can be seen that the function has a kink at $x = \xi$. Alternatively to Eq. (5.7) and under consideration of the graphical representation in Fig. 5.1, the fundamental solution can be expressed as:

$$W^*(x) = \begin{cases} -\dfrac{1}{2}(x - \xi) & \text{for } x \leq \xi \\ +\dfrac{1}{2}(x - \xi) & \text{for } x \geq \xi \end{cases} . \tag{5.8}$$

Fig. 5.1 Fundamental solution for a bar subjected to a unit concentrated load at location $x = \xi$: **a** schematic sketch of $W^*(x)$; **b** first order derivative; **c** second order derivative

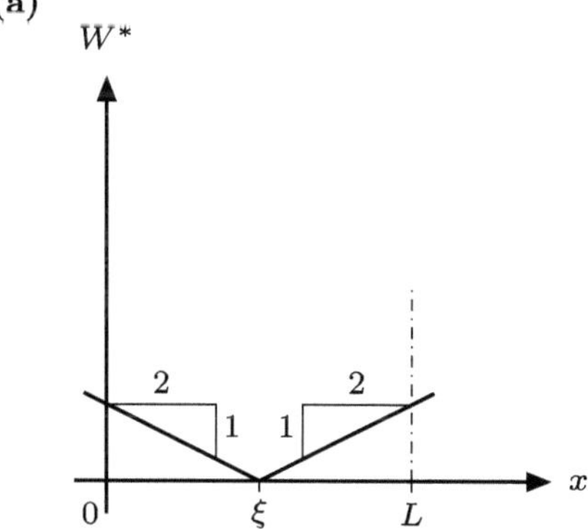

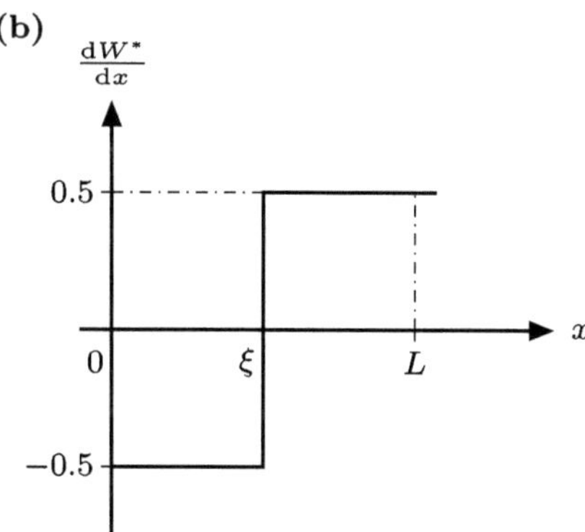

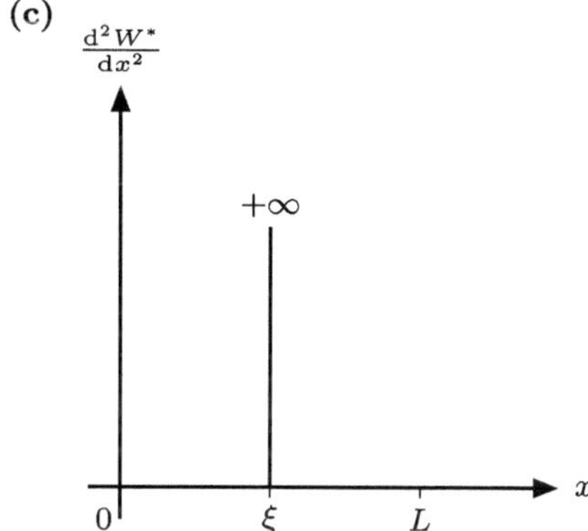

Furthermore, the boundary representation of the fundamental solution can be expressed in the following way:

$$W^*(x)\big|_0^L = \begin{cases} W^*|_0 = +\dfrac{1}{2}\xi & \text{for } x = 0 \\[2ex] W^*|_L = +\dfrac{1}{2}(L - \xi) & \text{for } x = L \end{cases}. \tag{5.9}$$

As a result of the course of $W^*(x)$, the first-order derivative shows a jump of magnitude 1, which corresponds to the introduced unit load $F(\xi) = 1$. The functional representation of the first-order derivative can be given as follows:

$$\frac{\mathrm{d}W^*(x)}{\mathrm{d}x} = \begin{cases} -\dfrac{1}{2} & \text{for } x < \xi \\[2ex] +\dfrac{1}{2} & \text{for } x > \xi \end{cases}, \tag{5.10}$$

or as the boundary representation:

$$\frac{\mathrm{d}W^*(x)}{\mathrm{d}x}\bigg|_0^L = \begin{cases} \dfrac{\mathrm{d}W^*(x)}{\mathrm{d}x}\bigg|_0 = -\dfrac{1}{2} & \text{for } x = 0 \\[2ex] \dfrac{\mathrm{d}W^*(x)}{\mathrm{d}x}\bigg|_L = +\dfrac{1}{2} & \text{for } x = L \end{cases}. \tag{5.11}$$

The second-order derivative is everywhere zero except for the load point where the value converges infinity.

Furthermore, it can be seen from Eq. (5.7) that the displacement of the infinite bar is constraint to zero at the load point: $W^*(x = \xi) = 0$. Alternatively to Eq. (5.10), the first-order derivative of Eq. (5.7) is given for $x \neq \xi$ as (cf. Fig. 5.1b)

$$\frac{\mathrm{d}W^*}{\mathrm{d}x} = \frac{1}{2} \times \mathrm{sgn}(x - \xi), \tag{5.12}$$

where 'sgn' is the signum function defined as:

$$\mathrm{sgn}(x - \xi) = \begin{cases} +1 & \text{for } x > \xi \\ 0 & \text{for } x = \xi \\ -1 & \text{for } x < \xi \end{cases}. \tag{5.13}$$

As can be seen from Fig. 5.1, the second-order derivative of Eq. (5.7) is given by the Dirac delta function (see Eq. (5.3)).

Inserting the fundamental solution (5.7) and boundary conditions,[1] i.e.

---

[1] The following refers now to the situation shown in Fig. 1.2b, i.e. a bar fixed at $x = 0$, loaded by a distributed load $p(x)$ and a point load F_0 at the right-hand boundary at $x = L$.

$$u_0 = u(x = 0) = 0, \tag{5.14}$$

$$N(x = L) = EA \frac{du}{dx} = F_0, \tag{5.15}$$

into the representation formula (5.6) gives

$$u(\xi) = -\int_0^L \frac{p}{EA} W^* dx - \underbrace{\frac{du}{dx}}_{F_0/(EA)} W^*\big|_L + \frac{du}{dx} W^*\big|_0 + u \frac{dW^*}{dx}\big|_L - \underbrace{u}_{0} \frac{dW^*}{dx}\big|_0$$

$$\tag{5.16}$$

or

$$u(\xi) = -\int_0^L \frac{p(x)}{EA} \frac{1}{2}|x - \xi|dx - \frac{F_0}{EA} \frac{1}{2}(L - \xi) + \left(\frac{du}{dx}\right)_0 \frac{1}{2}\xi + u_L \frac{1}{2}. \tag{5.17}$$

The last equation gives the displacement at any point ξ in the domain only in terms of boundary variables. The domain integral ($\int_0^L \cdots$) does not contain any unknown and can therefore be regarded as a constant term. To obtain two equations for the unknown boundary values $\left(\frac{du}{dx}\right)_0 = \left(\frac{du}{dx}\right)\big|_{(x=0)}$ and $u_L = u(x = L)$, the load point can be placed on the boundaries. With $\xi = 0$ and $\xi = L$ (see the graphical representation of the fundamental solution for these special cases in Fig. 5.2), the following two equations are obtained:

$$\underbrace{u(\xi = 0)}_{0} = -\int_0^L \frac{p(x)}{2EA} x dx - \frac{F_0 L}{2EA} + \frac{u_L}{2}, \tag{5.18}$$

$$u(\xi = L) = -\int_0^L \frac{p(x)}{2EA}(L - x)dx + \left(\frac{du}{dx}\right)_0 \frac{1}{2}L + \frac{u_L}{2}. \tag{5.19}$$

This system of two equations can be solved to obtain the two unknowns. It is assumed for simplicity that the distributed load is constant, i.e. $p(x) \rightarrow p_0 = $ const., to obtain the unknowns as:

$$u(x = L) = \frac{F_0 L}{EA} + \frac{p_0 L^2}{2EA}, \tag{5.20}$$

$$\left(\frac{du}{dx}\right)\Big|_{(x=0)} = \frac{F_0}{EA} + \frac{p_0 L}{EA}. \tag{5.21}$$

It should the noted here that the evaluation of the domain integral in Eq. (5.17) can be performed for $p(x) \rightarrow p_0 = $ const. as follows:

Fig. 5.2 Special cases of the fundamental solution at the boundaries: **a** $\xi = 0$; **b** $\xi = L$

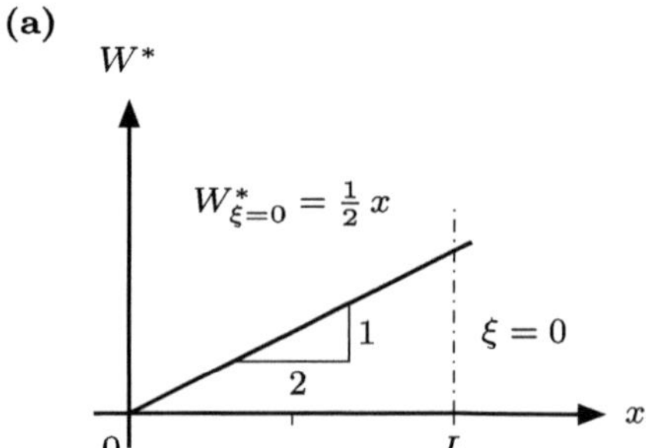

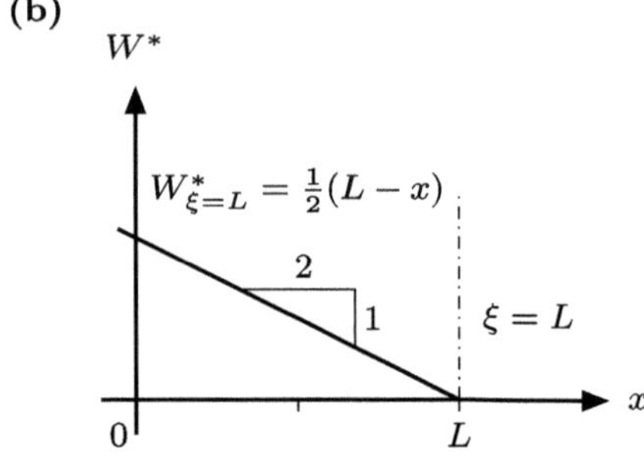

$$-\int_0^L \frac{p_0}{EA}\frac{1}{2}|x - \xi|\,dx = -\int_0^\xi \frac{p_0}{EA}\left\{-\frac{1}{2}(x - \xi)\right\}dx - \int_\xi^L \frac{p_0}{EA}\left\{+\frac{1}{2}(x - \xi)\right\}dx$$

$$= \frac{p_0}{2EA}\left[\frac{1}{2}x^2 - \xi x\right]_0^\xi - \frac{p_0}{2EA}\left[\frac{1}{2}x^2 - \xi x\right]_\xi^L$$

$$= \frac{p_0}{2EA}\left(-\xi^2 - \frac{1}{2}L^2 + \xi L\right). \tag{5.22}$$

The last step is the calculation of the value $u(\xi)$ inside the domain based on the boundary values. Inserting the two calculated boundary values in Eq. (5.17) gives the distribution of the displacement as:

$$u(\xi) = \frac{p_0}{EA}\left(L\xi - \frac{\xi^2}{2}\right) + \frac{F_0\xi}{EA}, \tag{5.23}$$

which is equal to the analytical solution, see [2]. However, this holds only for this simple case and is not true in general since only approximate solutions are obtained.

References

1. A.E.H. Love, *A Treatise on the Mathematical Theory of Elasticity* (Dover Publications, Mineola, 1944)
2. A. Öchsner, *Elasto-Plasticity of Frame Structure Elements: Modeling and Simulation of Rods and Beams* (Springer-Verlag, Berlin, 2014)
3. I. Sokolnikoff, *Mathematical theory of elasticity* (McGraw-Hill, New York, 1956)

Chapter 6
Comparison of the Methods

6.1 Example 1: Bar Loaded by a Distributed Load

Given is a cantilevered bar of length L with constant tensile stiffness EA as shown in Fig. 6.1. The bar is loaded by a constant distributed load p_0 in positive x-direction.

Use a finite element approach based on four elements, each of length $L/4$, and distinguish between linear and quadratic element formulations. Then use five domain nodes of equidistant spacing, i.e. $\Delta x = L/4$, for a finite difference approximation. Determine

- the horizontal displacement at the end of the bar, i.e. at $x = L$,
- the analytical solution and
- calculate the relative error between the analytical and the three numerical solutions.

6.1.1 Analytical Solution

The analytical solution is based on the differential equation in the form of Eq. (1.19). Considering that $p_x(x) \rightarrow p_0 = \text{const.}$ allows twice integrating the equation, i.e.,

$$EA\frac{\mathrm{d}^2 u_x(x)}{\mathrm{d}x^2} = -p_0 \,, \tag{6.1}$$

$$EA\frac{\mathrm{d}^1 u_x(x)}{\mathrm{d}x^1} = N_x(x) = -p_0 x + c_1 \,, \tag{6.2}$$

$$EA u_x(x) = -\frac{p_0 x^2}{2} + c_1 x + c_2 \,. \tag{6.3}$$

A. Öchsner, *An Introduction to the Classical Approximation Methods in Applied Mechanics*, SpringerBriefs in Computational Mechanics, https://doi.org/10.1007/978-3-032-06967-2_6

Fig. 6.1 Cantilevered tensile
bar loaded by a constant
distributed load

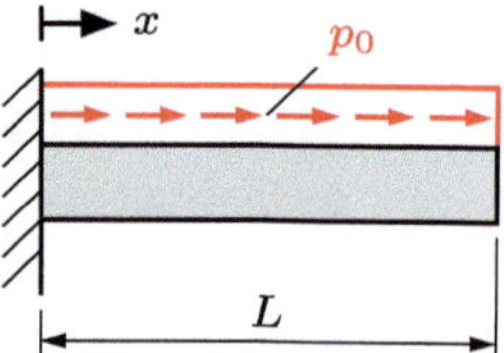

Fig. 6.2 Cutting free inside
the cantilever bar with
constant distributed load

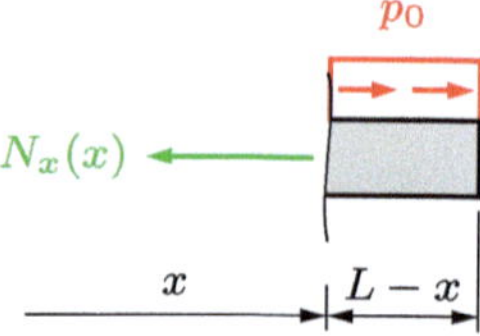

Fig. 6.3 Analytical solution
for the bar elongation along
the x-axis for the case of a
constant distributed load p_0

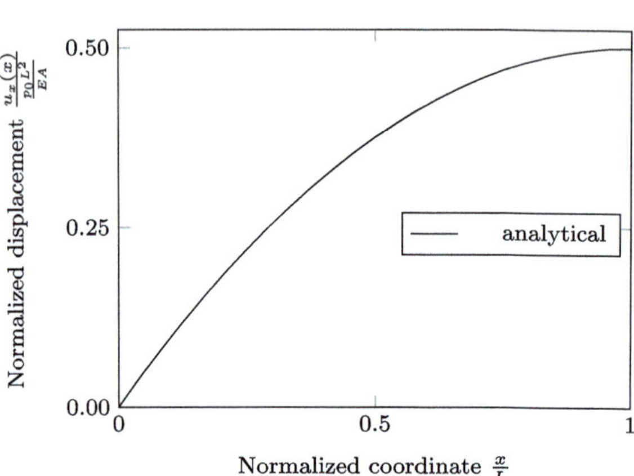

Consideration of the boundary conditions, i.e., $u_x(x = 0) = 0$ and $N_x(x = L) = 0$ (see Fig. 6.2 for $x \to L$, it follows that $(L - x) \to 0$), allows to determine the constants of integration as $c_2 = 0$ and $c_1 = p_0 L$.

Thus, the course of the displacement field is finally obtained as, see also Fig. 6.3:

$$u_x(x) = \frac{p_0 L^2}{EA} \left(-\frac{1}{2} \left(\frac{x}{L} \right)^2 + \left(\frac{x}{L} \right) \right). \tag{6.4}$$

The maximum displacement is obtained at the right-hand boundary ($x = L$) of the bar as:

$$u_x(L) = \frac{p_0 L^2}{2EA}. \tag{6.5}$$

It can be seen from Fig. 6.3 and Eq. (6.4) that a nonlinear, or more precisely a quadratic, displacement field is obtained.

6.1.2   Finite Element Approach

6.1.2.1   Four Linear Elements

The finite element discretization of the bar structure into four linear elements of equal length is shown in Fig. 6.4. This figure also includes the assigned node numbering $(1, \ldots, 5)$ as well as the element numbering $(\mathrm{I}, \ldots, \mathrm{IV})$.

The general stiffness matrix for a single linear element as provided in Eq. (3.22) can be adjusted to the problem at hand by setting $L \to \frac{L}{4}$ and assembling of the four elements to the global system of equations results in the following representation (for details on the procedure, see [3, 5]):

$$\frac{4EA}{L} \begin{bmatrix} 1 & -1 & 0 & 0 & 0 \\ -1 & 1+1 & -1 & 0 & 0 \\ 0 & -1 & 1+1 & -1 & 0 \\ 0 & 0 & -1 & 1+1 & -1 \\ 0 & 0 & 0 & -1 & 1 \end{bmatrix} \begin{bmatrix} u_1 \\ u_2 \\ u_3 \\ u_4 \\ u_5 \end{bmatrix} = \begin{bmatrix} \dfrac{p_0 L}{8} - R_1 \\ \dfrac{p_0 L}{8} + \dfrac{p_0 L}{8} \\ \dfrac{p_0 L}{8} + \dfrac{p_0 L}{8} \\ \dfrac{p_0 L}{8} + \dfrac{p_0 L}{8} \\ \dfrac{p_0 L}{8} \end{bmatrix}, \qquad (6.6)$$

or under consideration of the support condition at $x = 0$, i.e., $u_1 = 0$, as the reduced global system of equations:

$$\frac{4EA}{L} \begin{bmatrix} 2 & -1 & 0 & 0 \\ -1 & 2 & -1 & 0 \\ 0 & -1 & 2 & -1 \\ 0 & 0 & -1 & 1 \end{bmatrix} \begin{bmatrix} u_2 \\ u_3 \\ u_4 \\ u_5 \end{bmatrix} = \begin{bmatrix} \dfrac{p_0 L}{4} \\ \dfrac{p_0 L}{4} \\ \dfrac{p_0 L}{4} \\ \dfrac{p_0 L}{8} \end{bmatrix}. \qquad (6.7)$$

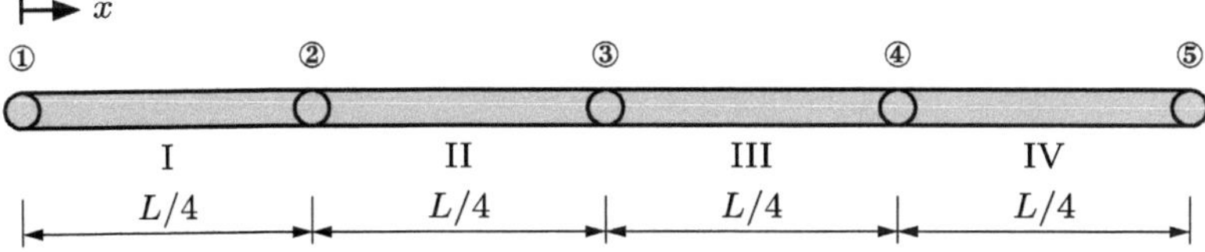

Fig. 6.4   Finite element discretization based on four linear bar elements

It should be noted here that Eq. (6.6) contains on the right-hand side the equivalent nodal loads (see Eq. (3.35)). Furthermore, it is assumed that the reaction force at node 1, i.e., R_1, acts in negative x-direction. The solution of this linear system of equations (the reduced formulation) can be obtained, for example, by inverting the stiffness matrix and multiplying with the load matrix, i.e., $\boldsymbol{u} = \boldsymbol{K}^{-1}\boldsymbol{f}$:

$$
\begin{bmatrix} u_2 \\ u_3 \\ u_4 \\ u_5 \end{bmatrix} = \frac{p_0 L^2}{EA} \begin{bmatrix} \dfrac{7}{32} \\ \dfrac{3}{8} \\ \dfrac{15}{32} \\ \dfrac{1}{2} \end{bmatrix} .
\tag{6.8}
$$

It can be seen from Eq. (6.8) that the maximum displacement at node 5 is equal to the analytical solution as given in Eq. (6.5). Thus, this finite element approach provides for this simple example the exact solution at the nodes without any error.

Let us look in the following on the entire displacement $u_x(x)$ field and not only on the nodal values. The exact displacement distribution can be taken from [4] as:

$$
u_x(x) = \frac{p_0 L^2}{EA} \left(\left(\frac{x}{L}\right)^1 - \frac{1}{2}\left(\frac{x}{L}\right)^2 \right) .
\tag{6.9}
$$

The finite element solution allows an element-wise representation of the displacement field according to Eq. (3.8). For the entire bar ($0 \le x \le L$), one may state:

$$
u_x(x) = \underbrace{\left(1 - \frac{x}{L/4}\right) u_1 + \left(\frac{x}{L/4}\right) u_2}_{0 \le x \le L/4}
\tag{6.10}
$$

$$
+ \underbrace{\left(1 - \frac{x - L/4}{L/4}\right) u_2 + \left(\frac{x - L/4}{L/4}\right) u_3}_{L/4 \le x \le L/2}
\tag{6.11}
$$

$$
+ \underbrace{\left(1 - \frac{x - 2L/4}{L/4}\right) u_3 + \left(\frac{x - 2L/4}{L/4}\right) u_4}_{L/2 \le x \le 3L/4}
\tag{6.12}
$$

$$
+ \underbrace{\left(1 - \frac{x - 3L/4}{L/4}\right) u_4 + \left(\frac{x - 3L/4}{L/4}\right) u_5}_{3L/4 \le x \le L} .
\tag{6.13}
$$

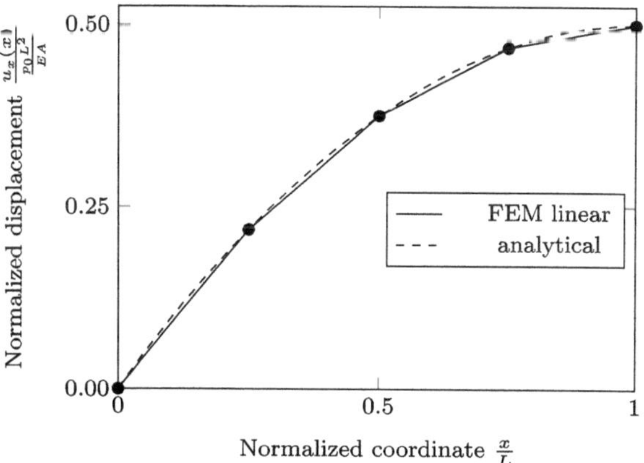

Fig. 6.5 Bar elongation along the x-axis based on four linear elements for the case of a constant distributed load p_0 (nodal values of the linear FEM approach represented by the marker •)

The graphical representation of the displacement field in Fig. 6.5 shows that the finite element solution is exact at all nodes. However, the course of the displacement field between two elemental nodes is linear for the finite element solution whereas the exact analytical solution provides a quadratic behavior.

6.1.2.2   Eight Linear Elements

The finite element discretization of the bar structure into eight linear elements of equal length is shown in Fig. 6.6. This figure also includes the assigned node numbering $(1, \ldots, 9)$ as well as the element numbering $(\mathrm{I}, \ldots, \mathrm{VIII})$.

The general stiffness matrix for a single linear element as provided in Eq. (3.22) can be adjusted to the problem at hand by setting $L \rightarrow \frac{L}{8}$ and assembling of the eight elements to the global system of equations results in the following representation (for details on the procedure, see [3, 5]):

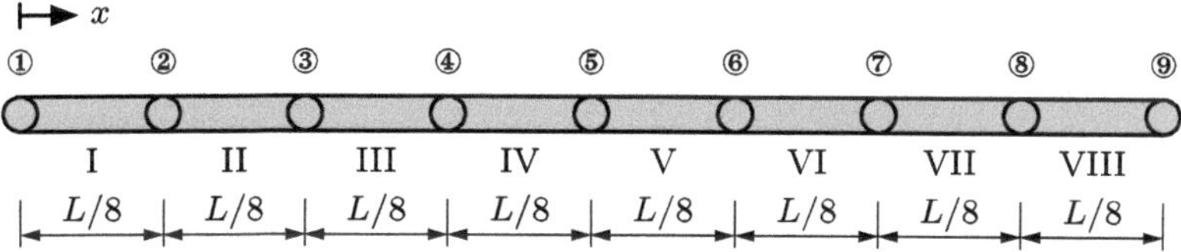

Fig. 6.6 Finite element discretization based on eight linear bar elements

$$\frac{8EA}{L}
\begin{bmatrix}
1 & -1 & 0 & 0 & 0 & 0 & 0 & 0 & 0 \\
-1 & 1+1 & -1 & 0 & 0 & 0 & 0 & 0 & 0 \\
0 & -1 & 1+1 & -1 & 0 & 0 & 0 & 0 & 0 \\
0 & 0 & -1 & 1+1 & -1 & 0 & 0 & 0 & 0 \\
0 & 0 & 0 & -1 & 1+1 & -1 & 0 & 0 & 0 \\
0 & 0 & 0 & 0 & -1 & 1+1 & -1 & 0 & 0 \\
0 & 0 & 0 & 0 & 0 & -1 & 1+1 & -1 & 0 \\
0 & 0 & 0 & 0 & 0 & 0 & -1 & 1+1 & -1 \\
0 & 0 & 0 & 0 & 0 & 0 & 0 & -1 & 1
\end{bmatrix}
\begin{bmatrix}
u_1 \\ u_2 \\ u_3 \\ u_4 \\ u_5 \\ u_6 \\ u_7 \\ u_8 \\ u_9
\end{bmatrix}
=
\begin{bmatrix}
\dfrac{p_0 L}{16} - R_1 \\
\dfrac{p_0 L}{16} + \dfrac{p_0 L}{16} \\
\dfrac{p_0 L}{16} + \dfrac{p_0 L}{16} \\
\dfrac{p_0 L}{16} + \dfrac{p_0 L}{16} \\
\dfrac{p_0 L}{16} + \dfrac{p_0 L}{16} \\
\dfrac{p_0 L}{16} + \dfrac{p_0 L}{16} \\
\dfrac{p_0 L}{16} + \dfrac{p_0 L}{16} \\
\dfrac{p_0 L}{16} + \dfrac{p_0 L}{16} \\
\dfrac{p_0 L}{16}
\end{bmatrix} , \quad (6.14)$$

or under consideration of the support condition at $x = 0$, i.e., $u_1 = 0$, as the reduced global system of equations:

$$\frac{8EA}{L}
\begin{bmatrix}
2 & -1 & 0 & 0 & 0 & 0 & 0 & 0 \\
-1 & 2 & -1 & 0 & 0 & 0 & 0 & 0 \\
0 & -1 & 2 & -1 & 0 & 0 & 0 & 0 \\
0 & 0 & -1 & 2 & -1 & 0 & 0 & 0 \\
0 & 0 & 0 & -1 & 2 & -1 & 0 & 0 \\
0 & 0 & 0 & 0 & -1 & 2 & -1 & 0 \\
0 & 0 & 0 & 0 & 0 & -1 & 2 & -1 \\
0 & 0 & 0 & 0 & 0 & 0 & -1 & 1
\end{bmatrix}
\begin{bmatrix}
u_2 \\ u_3 \\ u_4 \\ u_5 \\ u_6 \\ u_7 \\ u_8 \\ u_9
\end{bmatrix}
=
\begin{bmatrix}
\dfrac{p_0 L}{8} \\
\dfrac{p_0 L}{8} \\
\dfrac{p_0 L}{8} \\
\dfrac{p_0 L}{8} \\
\dfrac{p_0 L}{8} \\
\dfrac{p_0 L}{8} \\
\dfrac{p_0 L}{8} \\
\dfrac{p_0 L}{16}
\end{bmatrix} , \quad (6.15)$$

It should be noted here that Eq. (6.14) contains on the right-hand side the equivalent nodal loads (see Eq. (3.35)). Furthermore, it is assumed that the reaction force at node 1, i.e., R_1, acts in negative x-direction. The solution of this linear system of equations (the reduced formulation) can be obtained, for example, by inverting the stiffness matrix and multiplying with the load matrix, i.e., $\boldsymbol{u} = \boldsymbol{K}^{-1}\boldsymbol{f}$:

$$
\begin{bmatrix} u_2 \\ u_3 \\ u_4 \\ u_5 \\ u_6 \\ u_7 \\ u_8 \\ u_9 \end{bmatrix} = \frac{p_0 L^2}{EA} \begin{bmatrix} \dfrac{15}{128} \\[4pt] \dfrac{7}{32} \\[4pt] \dfrac{39}{128} \\[4pt] \dfrac{3}{8} \\[4pt] \dfrac{55}{128} \\[4pt] \dfrac{15}{32} \\[4pt] \dfrac{63}{128} \\[4pt] \dfrac{1}{2} \end{bmatrix} .
\tag{6.16}
$$

It can be seen from Eq. (6.16) that the maximum displacement at node 9 is again equal to the analytical solution as given in Eq. (6.5). Thus, this finite element approach provides for this simple example at the nodes the exact solution without any error.

Let us look in the following on the entire displacement $u_x(x)$ field and not only on the nodal values. The exact displacement distribution can be taken from [7] as:

$$
u_x(x) = \frac{p_0 L^2}{EA} \left(\left(\frac{x}{L}\right)^1 - \frac{1}{2}\left(\frac{x}{L}\right)^2 \right) .
\tag{6.17}
$$

The finite element solution allows an element-wise representation of the displacement field according to Eq. (3.8). For the entire bar ($0 \leq x \leq L$), one may state:

$$u_x(x) = \underbrace{\left(1 - \frac{x}{L/8}\right) u_1 + \left(\frac{x}{L/8}\right) u_2}_{0 \le x \le L/8} \tag{6.18}$$

$$+ \underbrace{\left(1 - \frac{x - L/8}{L/8}\right) u_2 + \left(\frac{x - L/8}{L/8}\right) u_3}_{L/8 \le x \le 2L/8} \tag{6.19}$$

$$+ \underbrace{\left(1 - \frac{x - 2L/8}{L/8}\right) u_3 + \left(\frac{x - 2L/8}{L/8}\right) u_4}_{2L/8 \le x \le 3L/8} \tag{6.20}$$

$$+ \underbrace{\left(1 - \frac{x - 3L/8}{L/8}\right) u_4 + \left(\frac{x - 3L/8}{L/8}\right) u_5}_{3L/8 \le x \le 4L/8} \tag{6.21}$$

$$+ \underbrace{\left(1 - \frac{x - 4L/8}{L/8}\right) u_4 + \left(\frac{x - 4L/8}{L/8}\right) u_5}_{4L/8 \le x \le 5L/8} \tag{6.22}$$

$$+ \underbrace{\left(1 - \frac{x - 5L/8}{L/8}\right) u_4 + \left(\frac{x - 5L/8}{L/8}\right) u_5}_{5L/8 \le x \le 6L/8} \tag{6.23}$$

$$+ \underbrace{\left(1 - \frac{x - 6L/8}{L/8}\right) u_4 + \left(\frac{x - 6L/8}{L/8}\right) u_5}_{6L/8 \le x \le 7L/8} \tag{6.24}$$

$$+ \underbrace{\left(1 - \frac{x - 7L/8}{L/8}\right) u_4 + \left(\frac{x - 7L/8}{L/8}\right) u_5}_{7L/8 \le x \le L} \, . \tag{6.25}$$

The graphical representation of the displacement field in Fig. 6.7 shows that the finite element solution is again exact at all nodes. Again, the course of the displacement field between two elemental nodes is linear for the finite element solution

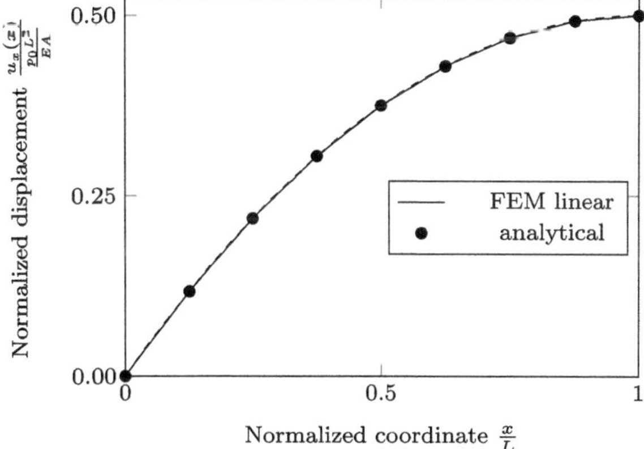

Fig. 6.7 Bar elongation along the x-axis based on eight linear elements for the case of a constant distributed load p_0 (nodal values of the linear FEM approach represented by the marker •)

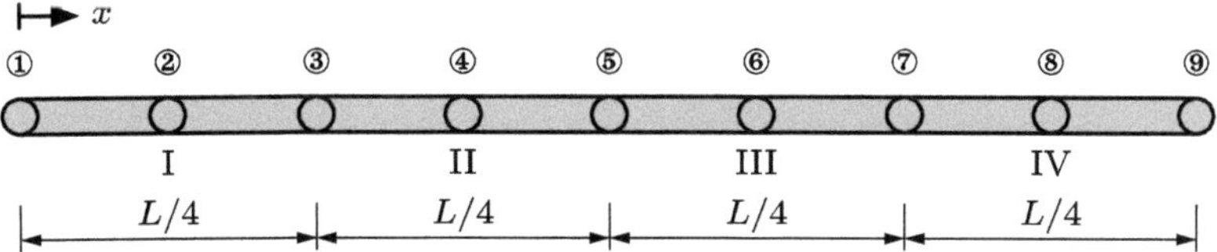

Fig. 6.8 Finite element discretization based on four quadratic bar elements

whereas the exact analytical solution provides a quadratic behavior. Nevertheless, the change from four to eight elements results in a much closer approximation of the exact distribution.

6.1.2.3 Four Quadratic Elements

The finite element discretization of the bar structure into four quadratic elements of equal length is shown in Fig. 6.8. This figure also includes the assigned node numbering $(1, \ldots, 9)$ as well as the element numbering $(\mathrm{I}, \ldots, \mathrm{IV})$.

The general stiffness matrix for a single quadratic element as provided in Eq. (3.41) can be adjusted to the problem at hand by setting $L \to \frac{L}{4}$ and assembling of the four elements to the global system of equations results in the following representation (for details on the procedure, see [3, 5]):

$$\frac{4EA}{3L}\begin{bmatrix} 7 & -8 & 1 & 0 & 0 & 0 & 0 & 0 & 0 \\ -8 & 16 & -8 & 0 & 0 & 0 & 0 & 0 & 0 \\ 1 & -8 & 7+7 & -8 & 1 & 0 & 0 & 0 & 0 \\ 0 & 0 & -8 & 16 & -8 & 0 & 0 & 0 & 0 \\ 0 & 0 & 1 & -8 & 7+7 & -8 & 1 & 0 & 0 \\ 0 & 0 & 0 & 0 & -8 & 16 & -8 & 0 & 0 \\ 0 & 0 & 0 & 0 & 1 & -8 & 7+7 & -8 & 1 \\ 0 & 0 & 0 & 0 & 0 & 0 & -8 & 16 & -8 \\ 0 & 0 & 0 & 0 & 0 & 0 & 1 & -8 & 7 \end{bmatrix} \begin{bmatrix} u_1 \\ u_2 \\ u_3 \\ u_4 \\ u_5 \\ u_6 \\ u_7 \\ u_8 \\ u_9 \end{bmatrix} = \begin{bmatrix} \dfrac{p_0 L}{24} - R_1 \\ \dfrac{p_0 L}{6} \\ \dfrac{p_0 L}{24} + \dfrac{p_0 L}{24} \\ \dfrac{p_0 L}{6} \\ \dfrac{p_0 L}{24} + \dfrac{p_0 L}{24} \\ \dfrac{p_0 L}{6} \\ \dfrac{p_0 L}{24} + \dfrac{p_0 L}{24} \\ \dfrac{p_0 L}{6} \\ \dfrac{p_0 L}{24} \end{bmatrix}, \tag{6.26}$$

or under consideration of the support condition at $x = 0$, i.e., $u_1 = 0$:

$$\frac{4EA}{3L}\begin{bmatrix} 16 & -8 & 0 & 0 & 0 & 0 & 0 & 0 \\ -8 & 14 & -8 & 1 & 0 & 0 & 0 & 0 \\ 0 & -8 & 16 & -8 & 0 & 0 & 0 & 0 \\ 0 & 1 & -8 & 14 & -8 & 1 & 0 & 0 \\ 0 & 0 & 0 & -8 & 16 & -8 & 0 & 0 \\ 0 & 0 & 0 & 1 & -8 & 14 & -8 & 1 \\ 0 & 0 & 0 & 0 & 0 & -8 & 16 & -8 \\ 0 & 0 & 0 & 0 & 0 & 1 & -8 & 7 \end{bmatrix} \begin{bmatrix} u_2 \\ u_3 \\ u_4 \\ u_5 \\ u_6 \\ u_7 \\ u_8 \\ u_9 \end{bmatrix} = \begin{bmatrix} \dfrac{p_0 L}{6} \\ \dfrac{p_0 L}{12} \\ \dfrac{p_0 L}{6} \\ \dfrac{p_0 L}{12} \\ \dfrac{p_0 L}{6} \\ \dfrac{p_0 L}{12} \\ \dfrac{p_0 L}{6} \\ \dfrac{p_0 L}{24} \end{bmatrix} . \tag{6.27}$$

Again, it should be noted here that Eq. (6.26) contains on the right-hand side the equivalent nodal loads (see Eq. (3.35)). Furthermore, it is assumed that the reaction force at node 1, i.e., R_1, acts in negative x-direction. The solution of this linear system of equations can be obtained, for example, by inverting the stiffness matrix and multiplying with the load matrix, i.e., $\boldsymbol{u} = \boldsymbol{K}^{-1}\boldsymbol{f}$:

$$
\begin{bmatrix} u_2 \\ u_3 \\ u_4 \\ u_5 \\ u_6 \\ u_7 \\ u_8 \\ u_9 \end{bmatrix} = \frac{p_0 L^2}{EA} \begin{bmatrix} \dfrac{15}{128} \\[2mm] \dfrac{7}{32} \\[2mm] \dfrac{39}{128} \\[2mm] \dfrac{3}{8} \\[2mm] \dfrac{55}{128} \\[2mm] \dfrac{15}{32} \\[2mm] \dfrac{63}{128} \\[2mm] \dfrac{1}{2} \end{bmatrix} .
\tag{6.28}
$$

Again, it can be seen from Eq. (6.28) that the maximum displacement at node 9 is equal to the analytical solution as given in Eq. (6.5). Thus, this quadratic finite element approach provides for this simple example the exact solution without any error.

Let us look in the following on the entire displacement $u_x(x)$ field and not only on the nodal values.

The finite element solution allows an element-wise representation of the displacement field according to Eq. (3.38). For the entire bar ($0 \leq x \leq L$), one may state:

$$
\begin{aligned}
u_x(x) = &\underbrace{\left(1 - 3\frac{x}{L/4} + 2\left(\frac{x}{L/4}\right)^2\right)u_1 + \left(4\frac{x}{L/4}\left(1 - \frac{x}{L/4}\right)\right)u_2 + \left(\frac{x}{L/4}\left(2\frac{x}{L/4} - 1\right)\right)u_3}_{0 \leq x \leq L/4} \\[2mm]
&+ \underbrace{\left(1 - 3\frac{x - \frac{1L}{4}}{L/4} + 2\left(\frac{x - \frac{1L}{4}}{L/4}\right)^2\right)u_3 + \left(4\frac{x - \frac{1L}{4}}{L/4}\left(1 - \frac{x - \frac{1L}{4}}{L/4}\right)\right)u_4 + \left(\frac{x - \frac{1L}{4}}{L/4}\left(2\frac{x - \frac{1L}{4}}{L/4} - 1\right)\right)u_5}_{L/4 \leq x \leq L/2} \\[2mm]
&+ \underbrace{\left(1 - 3\frac{x - \frac{2L}{4}}{L/4} + 2\left(\frac{x - \frac{2L}{4}}{L/4}\right)^2\right)u_5 + \left(4\frac{x - \frac{2L}{4}}{L/4}\left(1 - \frac{x - \frac{2L}{4}}{L/4}\right)\right)u_6 + \left(\frac{x - \frac{2L}{4}}{L/4}\left(2\frac{x - \frac{2L}{4}}{L/4} - 1\right)\right)u_7}_{L/2 \leq x \leq 3L/4} \\[2mm]
&+ \underbrace{\left(1 - 3\frac{x - \frac{3L}{4}}{L/4} + 2\left(\frac{x - \frac{3L}{4}}{L/4}\right)^2\right)u_7 + \left(4\frac{x - \frac{3L}{4}}{L/4}\left(1 - \frac{x - \frac{3L}{4}}{L/4}\right)\right)u_8 + \left(\frac{x - \frac{3L}{4}}{L/4}\left(2\frac{x - \frac{3L}{4}}{L/4} - 1\right)\right)u_9}_{3L/4 \leq x \leq L} .
\end{aligned}
\tag{6.29}
$$

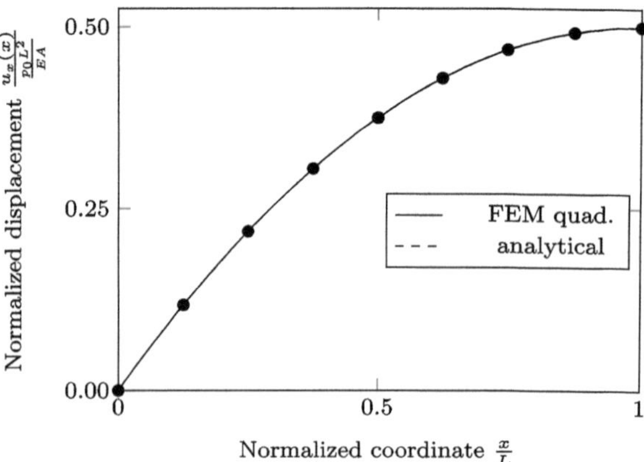

Fig. 6.9 Bar elongation along the x-axis for the case of a constant distributed load p_0 (nodal values of the quadratic FEM approach represented by the marker ●)

The graphical representation of the displacement field in Fig. 6.9 shows that the finite element solution is exact at all nodes. In addition, the course of the displacement field between three elemental nodes is now quadratic for the finite element solution as well as for the exact analytical solution.

6.1.3 Finite Difference Approach Based on Five Grid Points

The finite difference discretization of the bar structure based on five grid points of equal spacing is shown in Fig. 6.10. This figure also includes the assigned node numbering $(1, \ldots, 5)$.

The evaluation of the finite difference approximation according to Eq. (2.27) at the grid points without displacement boundary condition gives[1]:

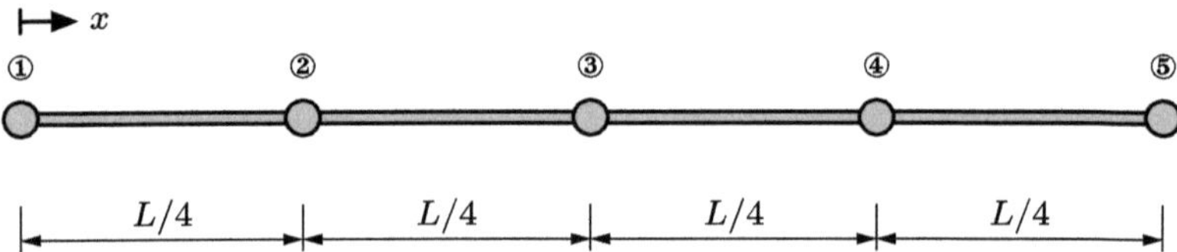

Fig. 6.10 Finite difference discretization based on five grid points

---

[1] We need to consider four equations for the four unknowns $u_2, \ldots, u_5$.

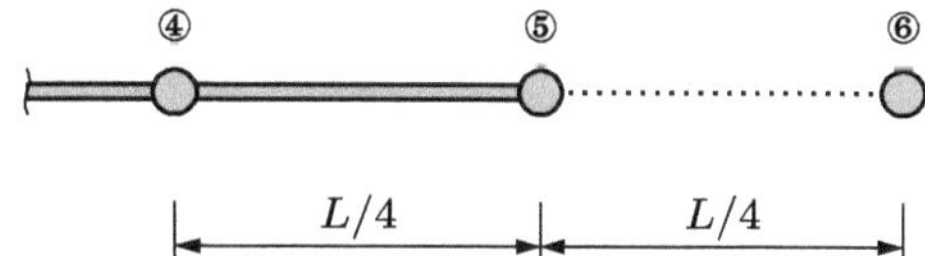

Fig. 6.11 Finite difference method: Fictitious node 6 outside the bar

$$\text{node 2:} \quad \frac{EA}{\Delta x}(-u_3 + 2u_2 - u_1) = p_0 \Delta x \,, \tag{6.30}$$

$$\text{node 3:} \quad \frac{EA}{\Delta x}(-u_4 + 2u_3 - u_2) = p_0 \Delta x \,, \tag{6.31}$$

$$\text{node 4:} \quad \frac{EA}{\Delta x}(-u_5 + 2u_4 - u_3) = p_0 \Delta x \,, \tag{6.32}$$

$$\text{node 5:} \quad \frac{EA}{\Delta x}(-u_6 + 2u_5 - u_4) = \frac{p_0 \Delta x}{2} \,. \tag{6.33}$$

Looking at Eq. (6.33), we can see that a non-existing node 6 is introduced. This node is called a fictitious node outside the bar and is schematically represented in Fig. 6.11.

The elimination of this fictitious node with its displacement u_6 requires further considerations. We use the relation for the internal normal force according to Eq. (1.18) and balance this internal reaction with the applied external force at node 5 (in our example, it holds that $F_5 = 0$):

$$EA\frac{\mathrm{d}u_x}{\mathrm{d}x}\bigg|_5 = N_5 \overset{!}{=} 0 \,. \tag{6.34}$$

To evaluate the gradient in Eq. (6.34), different formulations can be used, see [1, 2]. Using a centered difference scheme of higher accuracy ($O(\Delta x^2)$), the gradient at node 5 can be approximated as follows:

$$\frac{\mathrm{d}u_x}{\mathrm{d}x}\bigg|_5 \approx \frac{u_6 - u_4}{2\Delta x} \,. \tag{6.35}$$

Combining Eqs. (6.34) and (6.35), we get an additional condition to eliminate the fictitious displacement u_6:

$$u_6 = u_4 \,. \tag{6.36}$$

Introducing this condition and the support condition at node 1 ($u_1 = 0$) in the system of Eqs. (6.30)–(6.33), the final system of four equations for the four unknowns $u_2, \ldots, u_5$ is obtained:

$$\text{node 2:} \quad \frac{EA}{\Delta x}(2u_2 - u_3) = p_0 \Delta x, \tag{6.37}$$

$$\text{node 3:} \quad \frac{EA}{\Delta x}(-u_4 + 2u_3 - u_2) = p_0 \Delta x, \tag{6.38}$$

$$\text{node 4:} \quad \frac{EA}{\Delta x}(-u_5 + 2u_4 - u_3) = p_0 \Delta x, \tag{6.39}$$

$$\text{node 5:} \quad \frac{EA}{\Delta x}(-2u_4 + 2u_5) = \frac{p_0 \Delta x}{2}. \tag{6.40}$$

This set of four equations can be written in matrix form. Under consideration of $\Delta x = L/4$, one obtains the following formulation:

$$\frac{4EA}{L}\begin{bmatrix} 2 & -1 & 0 & 0 \\ -1 & 2 & -1 & 0 \\ 0 & -1 & 2 & -1 \\ 0 & 0 & -2 & 2 \end{bmatrix}\begin{bmatrix} u_2 \\ u_3 \\ u_4 \\ u_5 \end{bmatrix} = \begin{bmatrix} \dfrac{p_0 L}{4} \\ \dfrac{p_0 L}{4} \\ \dfrac{p_0 L}{4} \\ \dfrac{p_0 L}{8} \end{bmatrix}. \tag{6.41}$$

The solution of this linear system of equations can be obtained, for example, by inverting the coefficient matrix and multiplying with the matrix on the right hand side, i.e., $u = K^{-1}f$:

$$\begin{bmatrix} u_2 \\ u_3 \\ u_4 \\ u_5 \end{bmatrix} = \frac{p_0 L^2}{EA}\begin{bmatrix} \dfrac{13}{64} \\ \dfrac{11}{32} \\ \dfrac{27}{64} \\ \dfrac{7}{16} \end{bmatrix}. \tag{6.42}$$

It can be seen from Eq. (6.42) that the maximum displacement at node 5 is $u_5 = \frac{7p_0 L^2}{16EA} = 0.4375 \frac{p_0 L^2}{EA}$, whereas the analytical solution is given in Eq. (6.5) as $u_5 = \frac{p_0 L^2}{2EA}$. Thus, the finite difference approach provides a relative error of 12.5% in the maximum displacement. Let us look in the following on the entire displacement $u_x(x)$ field and not only on the grid point values.

The finite difference solution allows an interpolation between three grid points to represent the displacement field according to Eq. (2.17). For the entire bar ($0 \le x \le L$), one may state:

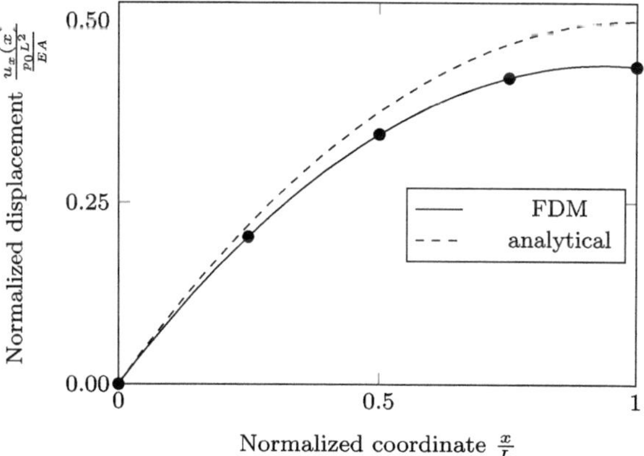

Fig. 6.12 Bar elongation along the x-axis for the case of a constant distributed load p_0 (nodal values of the FDM approach represented by the marker •)

$u_x(x) =$

$$\left(1 - 3\frac{x}{L/2} + 2\left(\frac{x}{L/2}\right)^2\right)u_1 + \left(4\frac{x}{L/2}\left(1 - \frac{x}{L/2}\right)\right)u_2 + \left(\frac{x}{L/2}\left(2\frac{x}{L/2} - 1\right)\right)u_3$$

$$\underbrace{\phantom{\left(1 - 3\frac{x}{L/2} + 2\left(\frac{x}{L/2}\right)^2\right)u_1 + \left(4\frac{x}{L/2}\left(1 - \frac{x}{L/2}\right)\right)u_2 + \left(\frac{x}{L/2}\left(2\frac{x}{L/2} - 1\right)\right)u_3}}_{0 \le x \le L/2}$$

$$+\left(1 - 3\frac{x - \frac{1L}{4}}{L/2} + 2\left(\frac{x - \frac{1L}{4}}{L/2}\right)^2\right)u_2 + \left(4\frac{x - \frac{1L}{4}}{L/2}\left(1 - \frac{x - \frac{1L}{4}}{L/2}\right)\right)u_3 + \left(\frac{x - \frac{1L}{4}}{L/2}\left(2\frac{x - \frac{1L}{4}}{L/2} - 1\right)\right)u_4$$

$$\underbrace{}_{L/4 \le x \le 3L/4}$$

$$+\left(1 - 3\frac{x - \frac{2L}{4}}{L/2} + 2\left(\frac{x - \frac{2L}{4}}{L/2}\right)^2\right)u_3 + \left(4\frac{x - \frac{2L}{4}}{L/2}\left(1 - \frac{x - \frac{2L}{4}}{L/2}\right)\right)u_4 + \left(\frac{x - \frac{2L}{4}}{L/2}\left(2\frac{x - \frac{2L}{4}}{L/2} - 1\right)\right)u_5 \ .$$

$$\underbrace{}_{L/2 \le x \le L}$$

The graphical representation of the displacement field in Fig. 6.12 shows that the finite difference solution and the exact solution have both a quadratic course but the values are different.

6.1.4   Boundary Element Approach Based on Two Boundary Nodes

The boundary element discretization of the bar structure based on two boundary nodes is shown in Fig. 6.13.

The boundary element solution of the displacement field can be obtained from Eq. (5.23) by setting $F_0 \to 0$:

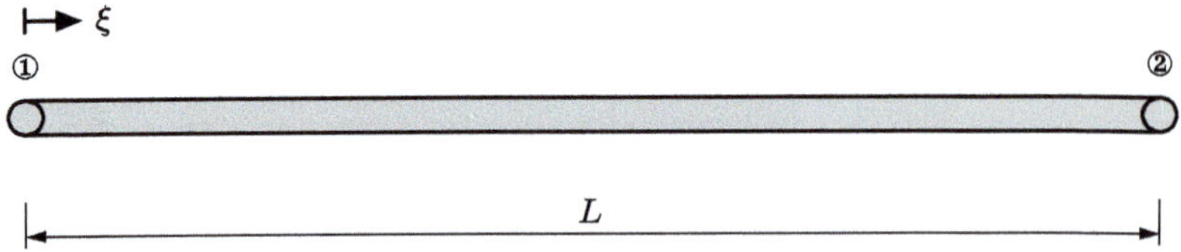

Fig. 6.13 Boundary element discretization based on two boundary nodes

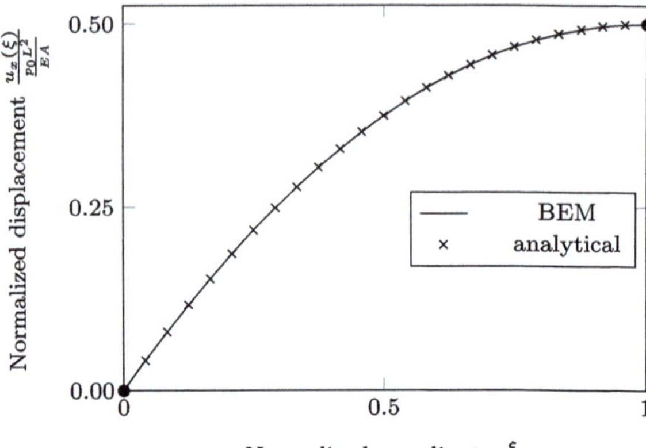

Fig. 6.14 Bar elongation along the x-axis for the case of a constant distributed load p_0 (nodal values of the BEM approach represented by the marker •)

Fig. 6.15 Cantilevered tensile bar loaded by a single force

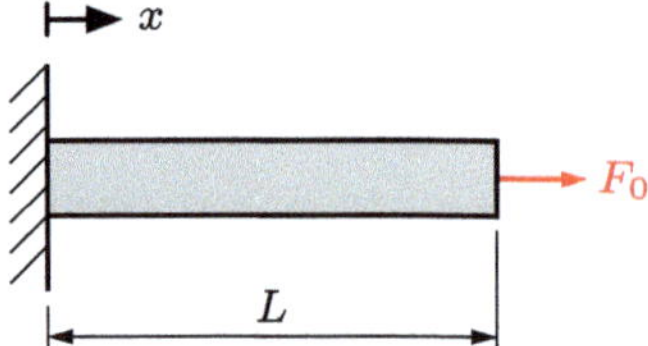

$$u(\xi) = \frac{p_0}{EA}\left(L\xi - \frac{\xi^2}{2}\right). \tag{6.43}$$

Comparing with the analytical solution in Eq. (6.4) and analyzing the graphical representation in Fig. 6.14, it can be concluded that the boundary solution is for this particular case of load and boundary conditions exact.

6.2 Example 2: Bar Loaded by a Single Force

Given is a cantilevered bar of length L with constant tensile stiffness EA as shown in Fig. 6.15. The bar is loaded by a single force F_0 at $x = L$ in positive x-direction.

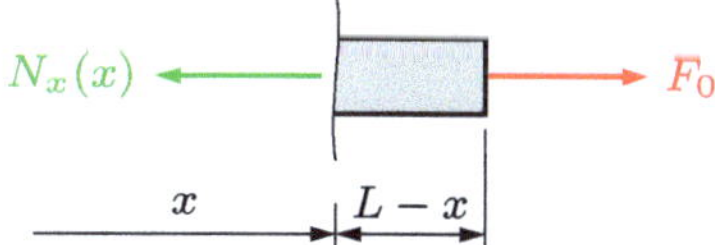

Fig. 6.16 Cutting free inside
the cantilever bar loaded by a
single force

6.2.1 Analytical Solution

The analytical solution is based on the differential equation in the form of Eq. (1.19).
Considering that $p_x(x) \rightarrow 0$ allows twice integrating the equation [7], i.e.,

$$EA\frac{\mathrm{d}^2 u_x(x)}{\mathrm{d}x^2} = 0, \tag{6.44}$$

$$EA\frac{\mathrm{d}^1 u_x(x)}{\mathrm{d}x^1} = N_x(x) = c_1, \tag{6.45}$$

$$EAu_x(x) = c_1 x + c_2. \tag{6.46}$$

Consideration of the boundary conditions, i.e., $u_x(x = 0) = 0$ and $N_x(x = L) = F_0$ (see Fig. 6.16), allows to determine the constants of integration as $c_2 = 0$ and $c_1 = F_0$.

Thus, the course of the displacement field is finally obtained as, see also Fig. 6.17:

$$u_x(x) = \frac{F_0 L}{EA} \times \left(\frac{x}{L}\right). \tag{6.47}$$

The maximum displacement is obtained at the right-hand boundary $(x = L)$ of
the bar as:

$$u_x(L) = \frac{F_0 L}{EA}. \tag{6.48}$$

It can be seen from Fig. 6.17 and Eq. (6.47) that a linear displacement field is
obtained.

6.2.2 Finite Element Approach

6.2.2.1 Four Linear Elements

The finite element discretization of the bar structure into four linear elements of equal
length is the same as shown in Fig. 6.4.

The general stiffness matrix for a single linear element as provided in Eq. (3.22)
can be adjusted to the problem at hand by setting $L \rightarrow \frac{L}{4}$ and assembling of the four

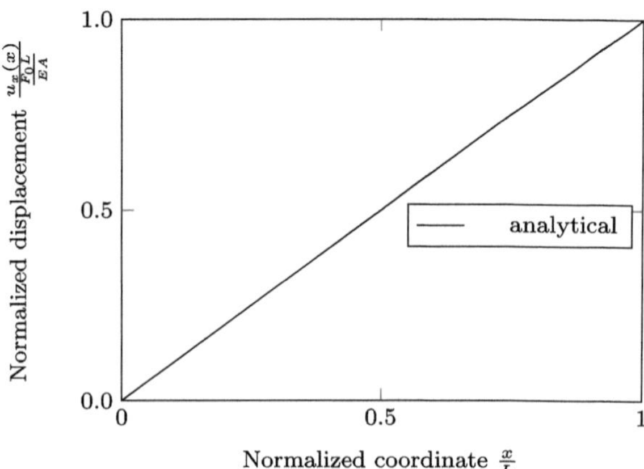

Fig. 6.17 Analytical solution for the bar elongation along the x-axis for the case of a point load F_0

elements to the global system of equations results in the following representation (for details on the procedure, see [3, 5]):

$$\frac{4EA}{L} \begin{bmatrix} 1 & -1 & 0 & 0 & 0 \\ -1 & 1+1 & -1 & 0 & 0 \\ 0 & -1 & 1+1 & -1 & 0 \\ 0 & 0 & -1 & 1+1 & -1 \\ 0 & 0 & 0 & -1 & 1 \end{bmatrix} \begin{bmatrix} u_1 \\ u_2 \\ u_3 \\ u_4 \\ u_5 \end{bmatrix} = \begin{bmatrix} -R_1 \\ 0 \\ 0 \\ 0 \\ F_0 \end{bmatrix}, \tag{6.49}$$

or under consideration of the support condition at $x = 0$, i.e., $u_1 = 0$, as the reduced global system of equations:

$$\frac{4EA}{L} \begin{bmatrix} 2 & -1 & 0 & 0 \\ -1 & 2 & -1 & 0 \\ 0 & -1 & 2 & -1 \\ 0 & 0 & -1 & 1 \end{bmatrix} \begin{bmatrix} u_2 \\ u_3 \\ u_4 \\ u_5 \end{bmatrix} = \begin{bmatrix} 0 \\ 0 \\ 0 \\ F_0 \end{bmatrix}. \tag{6.50}$$

It should be noted here that the reaction force at node 1, i.e., R_1, acts in negative x-direction. The solution of this linear system of equations (the reduced formulation) can be obtained, for example, by inverting the stiffness matrix and multiplying with the load matrix, i.e., $\boldsymbol{u} = \boldsymbol{K}^{-1}\boldsymbol{f}$:

$$
\begin{bmatrix} u_2 \\ u_3 \\ u_4 \\ u_5 \end{bmatrix} = \frac{F_0 L}{EA} \begin{bmatrix} \frac{1}{4} \\ \frac{1}{2} \\ \frac{3}{4} \\ 1 \end{bmatrix}.
\tag{6.51}
$$

It can be seen from Eq. (6.51) that the maximum displacement at node 5 is equal to the analytical solution as given in Eq. (6.48). Thus, this finite element approach provides for this simple example the exact solution at the nodes without any error.

Let us look in the following on the entire displacement $u_x(x)$ field and not only on the nodal values. The finite element solution allows an element-wise representation of the displacement field according to Eq. (3.8). For the entire bar ($0 \le x \le L$), one may state again:

$$
u_x(x) = \underbrace{\left(1 - \frac{x}{L/4}\right) u_1 + \left(\frac{x}{L/4}\right) u_2}_{0 \le x \le L/4}
\tag{6.52}
$$

$$
+ \underbrace{\left(1 - \frac{x - L/4}{L/4}\right) u_2 + \left(\frac{x - L/4}{L/4}\right) u_3}_{L/4 \le x \le L/2}
\tag{6.53}
$$

$$
+ \underbrace{\left(1 - \frac{x - 2L/4}{L/4}\right) u_3 + \left(\frac{x - 2L/4}{L/4}\right) u_4}_{L/2 \le x \le 3L/4}
\tag{6.54}
$$

$$
+ \underbrace{\left(1 - \frac{x - 3L/4}{L/4}\right) u_4 + \left(\frac{x - 3L/4}{L/4}\right) u_5}_{3L/4 \le x \le L}.
\tag{6.55}
$$

The graphical representation of the displacement field in Fig. 6.18 shows that the finite element solution is exact at all nodes. However, the course of the displacement field between two elemental nodes is linear for the finite element solution whereas the exact analytical solution provides a quadratic behavior.

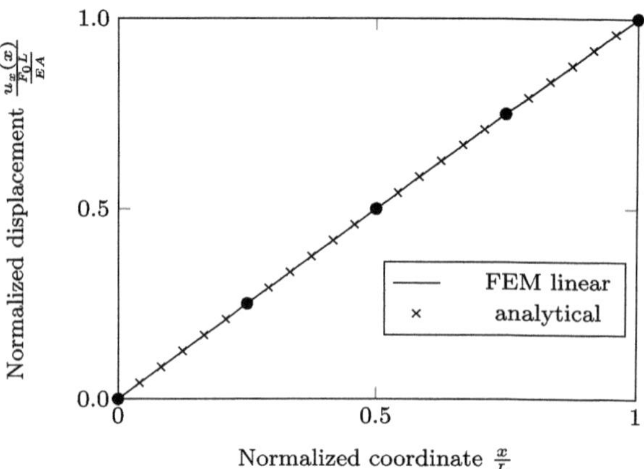

Fig. 6.18 Bar elongation along the x-axis based on four linear elements for the case of a point load F_0 (nodal values of the linear FEM approach represented by the marker ●)

The graphical representation of the displacement field in Fig. 6.18 shows that the finite element solution is in this case exact at all nodes, as well as exact in the course between the nodes.

6.2.2.2 Four Quadratic Elements

The finite element discretization of the bar structure into four quadratic elements of equal length is as shown in Fig. 6.8.

The general stiffness matrix for a single quadratic element as provided in Eq. (3.41) can be adjusted to the problem at hand by setting $L \rightarrow \frac{L}{4}$ and assembling of the four elements to the global system of equations results in the following representation (for details on the procedure, see [3, 5]):

$$\frac{4EA}{3L}\begin{bmatrix} 7 & -8 & 1 & 0 & 0 & 0 & 0 & 0 & 0 \\ -8 & 16 & -8 & 0 & 0 & 0 & 0 & 0 & 0 \\ 1 & -8 & 7+7 & -8 & 1 & 0 & 0 & 0 & 0 \\ 0 & 0 & -8 & 16 & -8 & 0 & 0 & 0 & 0 \\ 0 & 0 & 1 & -8 & 7+7 & -8 & 1 & 0 & 0 \\ 0 & 0 & 0 & 0 & -8 & 16 & -8 & 0 & 0 \\ 0 & 0 & 0 & 0 & 1 & -8 & 7+7 & -8 & 1 \\ 0 & 0 & 0 & 0 & 0 & 0 & -8 & 16 & -8 \\ 0 & 0 & 0 & 0 & 0 & 0 & 1 & -8 & 7 \end{bmatrix}\begin{bmatrix} u_1 \\ u_2 \\ u_3 \\ u_4 \\ u_5 \\ u_6 \\ u_7 \\ u_8 \\ u_9 \end{bmatrix} = \begin{bmatrix} -R_1 \\ 0 \\ 0 \\ 0 \\ 0 \\ 0 \\ 0 \\ 0 \\ F_0 \end{bmatrix}, \qquad (6.56)$$

or under consideration of the support condition at $x = 0$, i.e., $u_1 = 0$:

$$\frac{4EA}{3L}\begin{bmatrix} 16 & -8 & 0 & 0 & 0 & 0 & 0 & 0 \\ -8 & 14 & -8 & 1 & 0 & 0 & 0 & 0 \\ 0 & -8 & 16 & -8 & 0 & 0 & 0 & 0 \\ 0 & 1 & -8 & 14 & -8 & 1 & 0 & 0 \\ 0 & 0 & 0 & -8 & 16 & -8 & 0 & 0 \\ 0 & 0 & 0 & 1 & -8 & 14 & -8 & 1 \\ 0 & 0 & 0 & 0 & 0 & -8 & 16 & -8 \\ 0 & 0 & 0 & 0 & 0 & 1 & -8 & 7 \end{bmatrix}\begin{bmatrix} u_2 \\ u_3 \\ u_4 \\ u_5 \\ u_6 \\ u_7 \\ u_8 \\ u_9 \end{bmatrix} = \begin{bmatrix} 0 \\ 0 \\ 0 \\ 0 \\ 0 \\ 0 \\ 0 \\ F_0 \end{bmatrix}. \tag{6.57}$$

Again, it should be noted here the reaction force at node 1, i.e., R_1, acts in negative x-direction. The solution of this linear system of equations can be obtained, for example, by inverting the stiffness matrix and multiplying with the load matrix, i.e., $u = K^{-1}f$:

$$\begin{bmatrix} u_2 \\ u_3 \\ u_4 \\ u_5 \\ u_6 \\ u_7 \\ u_8 \\ u_9 \end{bmatrix} = \frac{F_0 L}{EA}\begin{bmatrix} \frac{1}{8} \\ \frac{1}{4} \\ \frac{3}{8} \\ \frac{1}{2} \\ \frac{5}{8} \\ \frac{3}{4} \\ \frac{7}{8} \\ 1 \end{bmatrix}. \tag{6.58}$$

Again, it can be seen from Eq. (6.58) that the maximum displacement at node 9 is equal to the analytical solution as given in Eq. (6.18). Thus, this quadratic finite element approach provides for this simple example the exact solution without any error.

Let us look in the following on the entire displacement $u_x(x)$ field and not only on the nodal values. Again, the finite element solution allows an element-wise representation of the displacement field according to Eq. (3.8). For the entire bar ($0 \leq x \leq L$), one may state:

$$u_x(x) = \underbrace{\left(1 - \frac{x}{L/8}\right) u_1 + \left(\frac{x}{L/8}\right) u_2}_{0 \leq x \leq L/8} \tag{6.59}$$

$$+ \underbrace{\left(1 - \frac{x - L/8}{L/8}\right) u_2 + \left(\frac{x - L/8}{L/8}\right) u_3}_{L/8 \leq x \leq 2L/8} \tag{6.60}$$

$$+ \underbrace{\left(1 - \frac{x - 2L/8}{L/8}\right) u_3 + \left(\frac{x - 2L/8}{L/8}\right) u_4}_{2L/8 \leq x \leq 3L/8} \tag{6.61}$$

$$+ \underbrace{\left(1 - \frac{x - 3L/8}{L/8}\right) u_4 + \left(\frac{x - 3L/8}{L/8}\right) u_5}_{3L/8 \leq x \leq 4L/8} \tag{6.62}$$

$$+ \underbrace{\left(1 - \frac{x - 4L/8}{L/8}\right) u_4 + \left(\frac{x - 4L/8}{L/8}\right) u_5}_{4L/8 \leq x \leq 5L/8} \tag{6.63}$$

$$+ \underbrace{\left(1 - \frac{x - 5L/8}{L/8}\right) u_4 + \left(\frac{x - 5L/8}{L/8}\right) u_5}_{5L/8 \leq x \leq 6L/8} \tag{6.64}$$

$$+ \underbrace{\left(1 - \frac{x - 6L/8}{L/8}\right) u_4 + \left(\frac{x - 6L/8}{L/8}\right) u_5}_{6L/8 \leq x \leq 7L/8} \tag{6.65}$$

$$+ \underbrace{\left(1 - \frac{x - 7L/8}{L/8}\right) u_4 + \left(\frac{x - 7L/8}{L/8}\right) u_5}_{7L/8 \leq x \leq L} . \tag{6.66}$$

The graphical representation of the displacement field in Fig. 6.19a shows that the finite element solution is again exact at all nodes. Again, the course of the displacement field between two elemental nodes is quadratic for the finite element solution whereas the exact analytical solution provides a linear behavior. Nevertheless, the

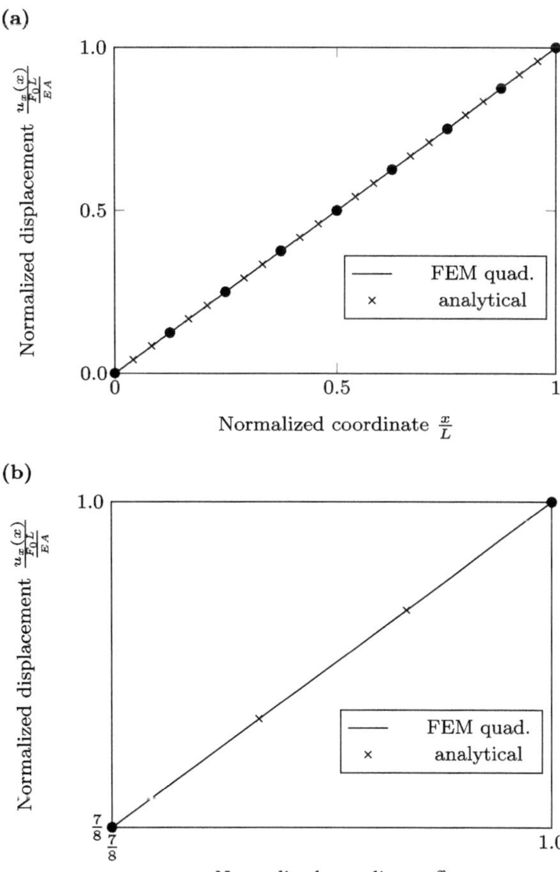

Fig. 6.19 Bar elongation along the x-axis based on eight linear elements for the case of a point load F_0 (nodal values of the linear FEM approach represented by the marker •): **a** general view and **b** magnification for element IV

difference between the quadratic and linear course is basically not visible, even in the magnification of Fig. 6.19b.

6.2.3 Finite Difference Approach Based on Five Grid Points

The finite difference discretization of the bar structure based on five grid points of equal spacing is as shown in Fig. 6.10.

The evaluation of the finite difference approximation according to Eq. (2.27) at the grid points without displacement boundary condition gives[2]:

---

[2] We need to consider four equations for the four unknowns $u_2, \ldots, u_5$.

$$\text{node 2:} \quad \frac{EA}{\Delta x}(-u_3 + 2u_2 - u_1) = 0\,, \tag{6.67}$$

$$\text{node 3:} \quad \frac{EA}{\Delta x}(-u_4 + 2u_3 - u_2) = 0\,, \tag{6.68}$$

$$\text{node 4:} \quad \frac{EA}{\Delta x}(-u_5 + 2u_4 - u_3) = 0\,, \tag{6.69}$$

$$\text{node 5:} \quad \frac{EA}{\Delta x}(-u_6 + 2u_5 - u_4) = 0\,. \tag{6.70}$$

Looking at Eq. (6.70), we can see that again a non-existing fictitious node 6 is introduced, see Fig. 6.11.

The elimination of this fictitious node with its displacement u_6 requires further considerations. We use the relation for the internal normal force according to Eq. (1.18) and balance this internal reaction with the applied external force at node 5:

$$EA\frac{\mathrm{d}u_x}{\mathrm{d}x}\bigg|_5 = N_5 \stackrel{!}{=} F_0\,. \tag{6.71}$$

To evaluate the gradient in Eq. (6.71), different formulations can be used, see [1, 2]. Using a centered difference scheme of higher accuracy ($O(\Delta x^2)$), the gradient at node 5 can be approximated as follows:

$$\frac{\mathrm{d}u_x}{\mathrm{d}x}\bigg|_5 \approx \frac{u_6 - u_4}{2\Delta x}\,. \tag{6.72}$$

Combining Eqs. (6.71) and (6.72), we get an additional condition to eliminate the fictitious displacement u_6:

$$u_6 = u_4 + \frac{2F_0\Delta x}{EA}\,. \tag{6.73}$$

Introducing this condition and the support condition at node 1 ($u_1 = 0$) in the system of Eqs. (6.67)–(6.70), the final system of four equations for the four unknowns $u_2, \ldots, u_5$ is obtained:

$$\text{node 2:} \quad \frac{EA}{\Delta x}(2u_2 - u_3) = 0\,, \tag{6.74}$$

$$\text{node 3:} \quad \frac{EA}{\Delta x}(-u_4 + 2u_3 - u_2) = 0\,, \tag{6.75}$$

$$\text{node 4:} \quad \frac{EA}{\Delta x}(-u_5 + 2u_4 - u_3) = 0\,, \tag{6.76}$$

$$\text{node 5:} \quad \frac{EA}{\Delta x}(-u_4 + u_5) = F_0\,. \tag{6.77}$$

This set of four equations can be written in matrix form. Under consideration of $\Delta x = L/4$, one obtains the following formulation:

$$\frac{4EA}{L}\begin{bmatrix} 2 & -1 & 0 & 0 \\ -1 & 2 & -1 & 0 \\ 0 & -1 & 2 & -1 \\ 0 & 0 & -1 & 1 \end{bmatrix}\begin{bmatrix} u_2 \\ u_3 \\ u_4 \\ u_5 \end{bmatrix} = \begin{bmatrix} 0 \\ 0 \\ 0 \\ F_0 \end{bmatrix}. \tag{6.78}$$

The solution of this linear system of equations can be obtained, for example, by inverting the coefficient matrix and multiplying with the matrix on the right hand side, i.e., $\boldsymbol{u} = \boldsymbol{K}^{-1}\boldsymbol{f}$:

$$\begin{bmatrix} u_2 \\ u_3 \\ u_4 \\ u_5 \end{bmatrix} = \frac{F_0 L}{EA}\begin{bmatrix} \frac{1}{4} \\ \frac{1}{2} \\ \frac{3}{4} \\ 1 \end{bmatrix}. \tag{6.79}$$

It can be seen from Eq. (6.79) that the maximum displacement at node 5 is equal to the analytical solution as is given in Eq. (6.5). Let us look in the following on the entire displacement $u_x(x)$ field and not only on the grid point values.

The finite difference solution allows an interpolation between three grid points to represent the displacement field according to Eq. (2.17). For the entire bar ($0 \leq x \leq L$), one may state:

$$u_x(x) =$$

$$\underbrace{\left(1 - 3\frac{x}{L/2} + 2\left(\frac{x}{L/2}\right)^2\right)u_1 + \left(4\frac{x}{L/2}\left(1 - \frac{x}{L/2}\right)\right)u_2 + \left(\frac{x}{L/2}\left(2\frac{x}{L/2} - 1\right)\right)u_3}_{0 \leq x \leq L/2}$$

$$+ \underbrace{\left(1 - 3\frac{x - \frac{1L}{4}}{L/2} + 2\left(\frac{x - \frac{1L}{4}}{L/2}\right)^2\right)u_2 + \left(4\frac{x - \frac{1L}{4}}{L/2}\left(1 - \frac{x - \frac{1L}{4}}{L/2}\right)\right)u_3 + \left(\frac{x - \frac{1L}{4}}{L/2}\left(2\frac{x - \frac{1L}{4}}{L/2} - 1\right)\right)u_4}_{L/4 \leq x \leq 3L/4}$$

$$+ \underbrace{\left(1 - 3\frac{x - \frac{2L}{4}}{L/2} + 2\left(\frac{x - \frac{2L}{4}}{L/2}\right)^2\right)u_3 + \left(4\frac{x - \frac{2L}{4}}{L/2}\left(1 - \frac{x - \frac{2L}{4}}{L/2}\right)\right)u_4 + \left(\frac{x - \frac{2L}{4}}{L/2}\left(2\frac{x - \frac{2L}{4}}{L/2} - 1\right)\right)u_5}_{L/2 \leq x \leq L}.$$

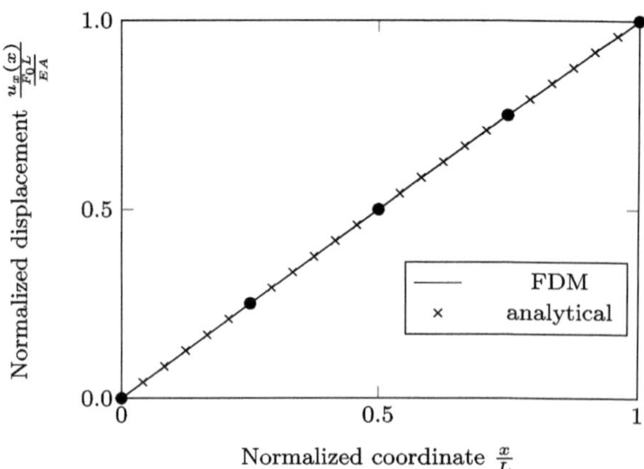

Fig. 6.20 Bar elongation along the x-axis for the case of a point load F_0 (nodal values of the FDM approach represented by the marker •)

The graphical representation of the displacement field in Fig. 6.20 shows that the finite difference solution and the exact solution are identical at the nodes and that the difference between the quadratic course (FDM) and the exact linear approach are not visible.

6.2.4   Boundary Element Approach Based on Two Boundary Nodes

The boundary element discretization of the bar structure based on two boundary nodes is as shown in Fig. 6.13.

The boundary element solution of the displacement field can be obtained from Eq. (5.23) by setting $p_0 \rightarrow 0$:

$$u(\xi) = \frac{F_0 \xi}{EA}. \tag{6.80}$$

Comparing with the analytical solution in Eq. (6.47) and analyzing the graphical representation in Fig. 6.21, it can be concluded that the boundary solution is for this particular case of load and boundary conditions exact.

6.3   Conclusions from the Examples

From the two examples presented in Sects. 6.1 and 6.2, the following conclusions can be drawn:

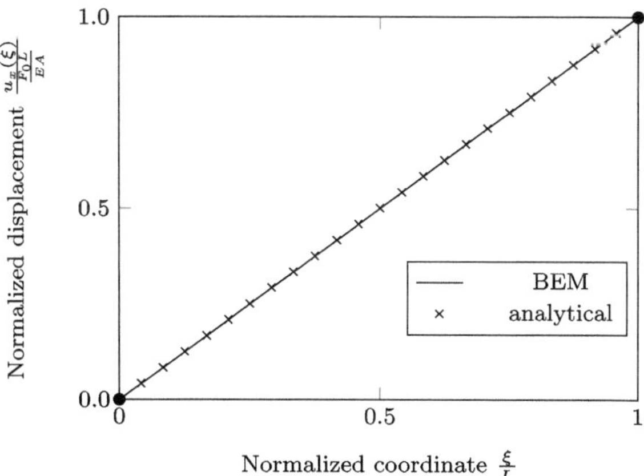

Fig. 6.21 Bar elongation along the x-axis for the case of a point load F_0 (nodal values of the BEM approach represented by the marker ●)

- Al the presented approximation methods are able to provide a solution for one-dimensional problems from the context of solid mechanics.
- Depending on the physical problem and the chosen approach, one may obtain from approximation methods for one-dimensional problems the exact analytical solution. However, for two- and three-dimensional problems, approximate methods provide, in general, solutions which are no longer exact.
- The accuracy can be many times increased by using more nodes and/or elements, i.e., a finer discretization of the domain.
- For the finite element method, higher-order elements may increase the accuracy. However, the physical nature of the problem should be considered.
- The one-dimensional approach has limitations for conclusions to the general multi-dimensional case. Here, it can be shown that some approximation methods reveal particular advantages and disadvantages: The finite difference method has difficulties in the case of arbitrary boundaries. The boundary element method is very suitable in the case of fracture mechanics and for semi-infinite domains. The finite volume method is common in the context of fluid mechanics.

References

1. K.-J. Bathe, *Finite Element Procedures* (Prentice-Hall, Upper Saddle River, 1996)
2. L. Collatz, *The Numerical Treatment of Differential Equations* (Springer-Verlag, Berlin, 1966)
3. A. Ochsner, *Elasto-Plasticity of Frame Structure Elements: Modeling and Simulation of Rods and Beams* (Springer-Verlag, Berlin 2014)
4. A. Ochsner, *Computational Statics and Dynamics: An Introduction Based on the Finite Element Method* (Springer, Singapore 2023)
5. Z. Javanbakht, A. Ochsner, *Computational Statics Revision Course* (Springer, Cham 2018)

Appendix
Mathematics

A.1 Integration by Parts

- One-dimensional case:

$$\int\limits_a^b f(x)\tfrac{\mathrm{d}g}{\mathrm{d}x}\,\mathrm{d}x = f(x)g(x)\big|_a^b - \int\limits_a^b \tfrac{\mathrm{d}f(x)}{\mathrm{d}x}g(x)\,\mathrm{d}x$$

$$= f(x)g(x)\big|_b - f(x)g(x)\big|_a - \int\limits_a^b \tfrac{\mathrm{d}f(x)}{\mathrm{d}x}g(x)\,\mathrm{d}x\,. \tag{A.1}$$

- Two-dimensional case (plane): $\mathrm{d}\Omega = \mathrm{d}A = \mathrm{d}x\mathrm{d}y$

$$\int\limits_\Omega f\tfrac{\mathrm{d}g}{\mathrm{d}x}\,\mathrm{d}\Omega = \int\limits_\Gamma fg\,n_x\,\mathrm{d}\Gamma - \int\limits_\Omega \tfrac{\mathrm{d}f}{\mathrm{d}x}g\,\mathrm{d}\Omega\,. \tag{A.2}$$

- Three-dimensional case (space): $\mathrm{d}\Omega = \mathrm{d}x\mathrm{d}y\mathrm{d}z$

$$\int\limits_\Omega f\tfrac{\mathrm{d}g}{\mathrm{d}x}\,\mathrm{d}\Omega = \int\limits_\Gamma fg\,n_x\,\mathrm{d}\Gamma - \int\limits_\Omega \tfrac{\mathrm{d}f}{\mathrm{d}x}g\,\mathrm{d}\Omega\,. \tag{A.3}$$

Remark: n_x is the cosine between the outward normal and the x-direction.

A. Öchsner, *An Introduction to the Classical Approximation Methods in Applied Mechanics*, SpringerBriefs in Computational Mechanics, https://doi.org/10.1007/978-3-032-06967-2

Or more general (Ω is the domain and Γ refers to the boundary):

$$\int_{\Omega} f g_{,i}\, \mathrm{d}\Omega = \int_{\Gamma} f g\, n_i\, \mathrm{d}\Gamma - \int_{\Omega} f_{,i}\, g\, \mathrm{d}\Omega\,, \tag{A.4}$$

$$\int_{\Omega} f_{,i}\, g\, \mathrm{d}\Omega = \int_{\Gamma} f g\, n_i\, \mathrm{d}\Gamma - \int_{\Omega} f g_{,i}\, \mathrm{d}\Omega\,. \tag{A.5}$$

- Green-Gauss theorem [2]: α: scalar function; $\boldsymbol{b}$: vector function; $\nabla = \left[\frac{\partial}{\partial x}\ \frac{\partial}{\partial y}\ \frac{\partial}{\partial z}\right]^{\mathrm{T}}$: Nabla operator.

$$\int_{\Omega} \alpha \nabla^{\mathrm{T}} \boldsymbol{b}\, \mathrm{d}\Omega = \int_{\Gamma} \alpha \boldsymbol{b}^{\mathrm{T}} \boldsymbol{n}\, \mathrm{d}\Gamma - \int_{\Omega} (\nabla^{\mathrm{T}}\alpha)\boldsymbol{b}\, \mathrm{d}\Omega\,, \tag{A.6}$$

where $\boldsymbol{n}$ is the unit outward vector acting on the boundary surface.

- Special case: $\boldsymbol{b} \to \beta \boldsymbol{b}$ (β: scalar)

$$\int_{\Omega} \alpha \nabla^{\mathrm{T}} (\beta \boldsymbol{b})\, \mathrm{d}\Omega = \int_{\Gamma} \alpha (\beta \boldsymbol{b})^{\mathrm{T}} \boldsymbol{n}\, \mathrm{d}\Gamma - \int_{\Omega} (\nabla^{\mathrm{T}}\alpha)(\beta \boldsymbol{b})\, \mathrm{d}\Omega\,. \tag{A.7}$$

- Special case: $\boldsymbol{b} \to \boldsymbol{K} \boldsymbol{b}$ ($\boldsymbol{K}$: matrix)

$$\int_{\Omega} \alpha \nabla^{\mathrm{T}} (\boldsymbol{K} \boldsymbol{b})\, \mathrm{d}\Omega = \int_{\Gamma} \alpha (\boldsymbol{K} \boldsymbol{b})^{\mathrm{T}} \boldsymbol{n}\, \mathrm{d}\Gamma - \int_{\Omega} (\nabla^{\mathrm{T}}\alpha)(\boldsymbol{K} \boldsymbol{b})\, \mathrm{d}\Omega\,. \tag{A.8}$$

In the notation of the weighted residual method, the following formulation is convenient:

$$\int_{V} \boldsymbol{W}^{\mathrm{T}} \boldsymbol{\mathcal{L}}_1^{\mathrm{T}} (\boldsymbol{C}\boldsymbol{\mathcal{L}}_1 \boldsymbol{u})\, \mathrm{d}V = \int_{A} \boldsymbol{W}^{\mathrm{T}} (\boldsymbol{C}\boldsymbol{\mathcal{L}}_1 \boldsymbol{u})^{\mathrm{T}} \boldsymbol{n}\, \mathrm{d}A - \int_{V} (\boldsymbol{\mathcal{L}}_1 \boldsymbol{W})^{\mathrm{T}} (\boldsymbol{C}\boldsymbol{\mathcal{L}}_1 \boldsymbol{u})\, \mathrm{d}V\,. \tag{A.9}$$

A.1 Example: One-dimensional integration by parts

Calculate the definite integral of

$$\int_{0}^{1} x^2 \mathrm{e}^{x}\, \mathrm{d}x\,. \tag{A.10}$$

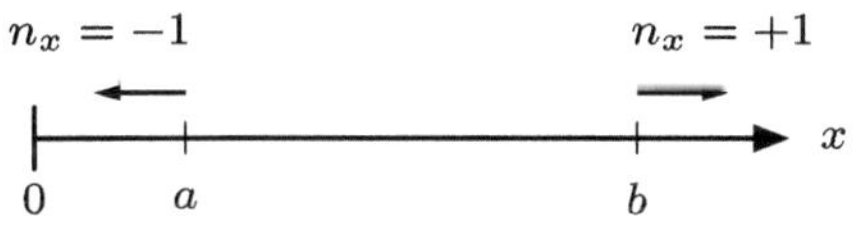

Fig. A.1 Domain and boundary representation for a one-dimensional problem

A.1 **Solution**

$$\int_0^1 \underbrace{x^2}_{f}\, \underbrace{e^x}_{g'}\, \mathrm{d}x = \left[x^2 e^x\right]_0^1 - \int_0^1 2x e^x \mathrm{d}x \,. \tag{A.11}$$

With

$$\int_0^1 \underbrace{x}_{f}\, \underbrace{e^x}_{g'}\, \mathrm{d}x = \left[x e^x\right]_0^1 - \int_0^1 1 e^x \mathrm{d}x \,. \tag{A.12}$$

$$\int_0^1 x^2 e^x \mathrm{d}x = \left[x^2 e^x - 2x e^x\right]_0^1 + 2\int_0^1 e^x \mathrm{d}x$$

$$= e - 2e + 2\left[e^x\right]_0^1 = e - 2 \approx 0.71828\,. \tag{A.13}$$

A.2 **Example: Simplification of the Green-Gauss Theorem**

Simplify the GREEN-GAUSS theorem given in Eq. (A.6) to the one-dimensional case given in Eq. (A.1).

A.2 **Solution**

The consideration of the simplifications $\nabla^{\mathrm{T}} \to \frac{\mathrm{d}\cdots}{\mathrm{d}x}$, $\boldsymbol{\alpha} \to \alpha(x)$ and $\boldsymbol{b} \to b(x)$ in Eq. (A.6) gives:

$$\int_\Omega \alpha(x)\frac{\mathrm{d}b(x)}{\mathrm{d}x}\,\mathrm{d}\Omega = \int_\Gamma \alpha(x)b(x)n_x\,\mathrm{d}\Gamma - \int_\Omega \frac{\mathrm{d}\alpha(x)}{\mathrm{d}x}b(x)\,\mathrm{d}x \,. \tag{A.14}$$

Assuming a domain $\Omega = [a, b]$ with boundaries at $x = a$ and $x = b$, the normal vectors at the boundaries result as $n_x|_a = -1$ and $n_x|_b = +1$, see Fig. A.1.

Thus, Eq. (A.14) can be transformed into:

$$\int_a^b \alpha(x)\frac{\mathrm{d}b(x)}{\mathrm{d}x}\,\mathrm{d}\Omega = \alpha(x)b(x)(+1)|_b + \alpha(x)b(x)(-1)|_a - \int_a^b \frac{\mathrm{d}\alpha(x)}{\mathrm{d}x}b(x)\,\mathrm{d}x \,.$$

$$\tag{A.15}$$

It should be noted here that the evaluation of the boundary integral in Eq. (A.14) requires the evaluation of this integral at all boundaries and summing up the obtained values.

A.2 Integration and Coordinate Transformation

(a) One-dimensional case:

Let $T : \mathbb{R} \to \mathbb{R}$ given by $x = g(u)$ be a one-dimensional transformation from S to R. If g has a continuous partial derivative such that the Jacobian determinant [1] is never zero, then

$$\int_R f(x)\mathrm{d}x = \int_S f(g(u)) \left| \frac{\mathrm{d}x}{\mathrm{d}u} \right| \mathrm{d}u \,, \tag{A.16}$$

where the Jacobian determinant[1] is $J = \left| \frac{\mathrm{d}x}{\mathrm{d}u} \right| = \frac{\mathrm{d}x}{\mathrm{d}u} = x_u$. Pay attention to the fact that the symbol $|\ldots|$ represents the determinant and should not be confused with the absolute value.

(b) Two-dimensional case:

Let $T : \mathbb{R}^2 \to \mathbb{R}^2$ given by $x = g(u, v)$ and $y = h(u, v)$ be a transformation on the plane that is one from a region S to a region R. If g and h have continuous partial derivatives such that the Jacobian determinant is never zero, then

$$\iint_R f(x, y)\mathrm{d}y\mathrm{d}x = \iint_S f(g(u, v), h(u, v)) \left| \frac{\partial(x, y)}{\partial(u, v)} \right| \mathrm{d}u\,\mathrm{d}v \,, \tag{A.17}$$

where the Jacobian determinant is $J = \left| \dfrac{\partial(x, y)}{\partial(u, v)} \right| = \left| \begin{matrix} x_u & x_v \\ y_u & y_v \end{matrix} \right| = x_u \cdot y_v - x_v \cdot y_u = \dfrac{\partial x}{\partial u} \cdot \dfrac{\partial y}{\partial v} - \dfrac{\partial x}{\partial v} \cdot \dfrac{\partial y}{\partial u}.$

(c) Three-dimensional case:

Let $T : \mathbb{R}^3 \to \mathbb{R}^3$ given by $x = g(u, v, w)$, $y = h(u, v, w)$ and $z = k(u, v, w)$ be a transformation on the space that is one from a space S to a space R. If g, h and k have continuous partial derivatives such that the Jacobian determinant is never zero, then

---

[1] A scalar a can be viewed as a 1×1 matrix and the corresponding determinant is just the number a itself.

$$\iiint\limits_{R} f(x, y, z)\,dz\,dy\,dx =$$

$$\iiint\limits_{S} f(g(u, v, w), h(u, v, w), k(u, v, w)) \left| \frac{\partial(x, y, z)}{\partial(u, v, w)} \right| dw\,du\,dv \,,$$

$$(A.18)$$

where the Jacobian determinant is $J = \left| \dfrac{\partial(x, y, z)}{\partial(u, v, w)} \right| = \begin{vmatrix} x_u & x_v & x_w \\ y_u & y_v & y_w \\ z_u & z_v & z_w \end{vmatrix}$.

Remark: A useful fact is that the Jacobian determinant of the inverse transformation is the reciprocal of the Jacobian determinant of the original transformation, e. g.

$$\left| \frac{\partial(x, y)}{\partial(u, v)} \right| = \left| \frac{\partial(u, v)}{\partial(x, y)} \right|^{-1} \,, \tag{A.19}$$

or

$$\left| \frac{\partial(u, v)}{\partial(x, y)} \right| = \left| \frac{\partial(x, y)}{\partial(u, v)} \right|^{-1} \,. \tag{A.20}$$

A useful relation in the scope of coordinate transformations holds for the inverse Jacobian matrix together with its determinant in the following way:

$$\underbrace{\begin{bmatrix} u_x & u_y & u_z \\ v_x & v_y & v_z \\ w_x & w_y & w_z \end{bmatrix}}_{J} = \underbrace{\begin{bmatrix} x_u & x_v & x_w \\ y_u & y_v & y_w \\ z_u & z_v & z_w \end{bmatrix}^{-1}}_{J'^{-1}} = \frac{1}{J} \cdot \begin{bmatrix} y_v z_w - y_w z_v & -x_v z_w + x_w z_v & x_v y_w - x_w y_v \\ -y_u z_w + y_w z_u & x_u z_w - x_w z_u & -x_u y_w + x_w y_u \\ y_u z_v - y_v z_u & -x_u z_v + x_v z_u & x_u y_v - x_v y_u \end{bmatrix} \,,$$

$$(A.21)$$

where the Jacobian determinant J is given by

$$J = \left| \frac{\partial(x, y, z)}{\partial(u, v, w)} \right| = x_u y_v z_w + x_v y_w z_u + x_w y_u z_v - x_v y_u z_w - x_w y_v z_u - x_u y_w z_v \,. \tag{A.22}$$

For the 2D case, Eq. (A.21) can be simplified to

$$\underbrace{\begin{bmatrix} u_x & u_y \\ v_x & v_y \end{bmatrix}}_{J} = \underbrace{\begin{bmatrix} x_u & x_v \\ y_u & y_v \end{bmatrix}^{-1}}_{J'^{-1}} = \frac{1}{J} \cdot \begin{bmatrix} y_v & -x_v \\ -y_u & x_u \end{bmatrix} \,, \tag{A.23}$$

where the Jacobian determinant J is given by

$$J = \left| \frac{\partial(x, y)}{\partial(u, v)} \right| = \left| \begin{matrix} x_u & x_v \\ y_u & y_v \end{matrix} \right| = x_u y_v - x_v y_u \ . \tag{A.24}$$

For the 1D case, Eq. (A.21) can be simplified to

$$\underbrace{\left[u_x \right]}_{J} = \underbrace{\left[x_u \right]^{-1}}_{J'^{-1}} = \frac{1}{J}, \tag{A.25}$$

where the Jacobian determinant J is given by

$$J = \left| \frac{\mathrm{d}x}{\mathrm{d}u} \right| = x_u \ . \tag{A.26}$$

A.3 Example: One-dimensional integration and coordinate transformation for a constant

Given is the integral $\int_0^L \frac{EA}{L^2} \times 1 \mathrm{d}x$. Change the variable from $0 \le x \le L$ to $-1 \le \xi \le 1$ and calculate the definite integral.

A.3 Solution

The transformation between the variables can be written as $\xi = \frac{2x}{L} - 1$ or as $\mathrm{d}\xi = \frac{2}{L}\mathrm{d}x$. Thus,

$$\int_0^L \frac{EA}{L^2} \times 1 \, \mathrm{d}x = \frac{EA}{L^2} \int_0^L 1 \, \mathrm{d}x = \frac{EA}{L^2} \int_{-1}^1 1 \times \frac{L}{2} \, \mathrm{d}\xi$$

$$= \frac{EA}{L^2} \times \frac{L}{2} [\xi]_{-1}^1 = \frac{EA}{2L}(1 + 1) = \frac{EA}{L}. \tag{A.27}$$

A.4 Example: One-dimensional integration and coordinate transformation for a linear function

Given is the integral $\int_0^L \frac{EA}{L^2} \times x \, \mathrm{d}x$. Change the variable from $0 \le x \le L$ to $-1 \le \xi \le 1$ and calculate the definite integral.

A.4 Solution

The transformation between the variables can be written as $\xi = \frac{2x}{L} - 1$ or as $\mathrm{d}\xi = \frac{2}{L}\mathrm{d}x$.
Solution without transformation:

$$\int_0^L \frac{EA}{L^2} \times x \, \mathrm{d}x = \frac{EA}{L^2} \left[\frac{x^2}{2} \right]_0^L = \frac{EA}{L^2} \times \frac{L^2}{2} = \frac{EA}{2}. \tag{A.28}$$

Solution with transformation:

$$\int_0^L \frac{EA}{L^2} \times x \, dx = \frac{EA}{L^2} \int_0^L x \, dx = \frac{EA}{L^2} \int_{-1}^1 \underbrace{\frac{L}{2}(\xi+1)}_{x} \times \frac{L}{2} \, d\xi$$

$$= \frac{EAL^2}{L^2 \, 4}\left[\frac{\xi^2}{2}+\xi\right]_{-1}^1 = \frac{EA}{4}\left\{(\tfrac{1}{2}+1)-(\tfrac{1}{2}-1)\right\} = \frac{EA}{2}. \qquad (A.29)$$

A.5 Example: Two-dimensional integration and coordinate transformation

Given is the integral $\int_{-a}^a \int_{-b}^b x^2 y^2 dx dy$. Change the variables from $-a \le x \le a$ and $-b \le y \le b$ to $-1 \le \xi \le 1$ and $-1 \le \eta \le 1$ for the transformations $\xi = \frac{x}{a}$ and $\eta = \frac{y}{b}$. Calculate the definite integral.

A.5 Solution

Solution without transformation:

$$\int_{-a}^a \int_{-b}^b x^2 y^2 dx dy = \int_{-a}^a x^2 \left[\frac{y^3}{3}\right]_{-b}^b dx = \int_{-a}^a x^2 \left(\frac{b^3}{3}+\frac{b^3}{3}\right) dx$$

$$= \frac{2b^3}{3}\int_{-a}^a x^2 dx = \frac{2b^3}{3}\left[\frac{x^3}{3}\right]_{-a}^a = \frac{4}{9}a^3 b^3 . \qquad (A.30)$$

Solution with transformation:

$$J = \frac{\partial x}{\partial \xi}\frac{\partial y}{\partial \eta} - \frac{\partial x}{\partial \eta}\frac{\partial y}{\partial \xi} = ab - 0 . \qquad (A.31)$$

$$\int_{-1}^1 \int_{-1}^1 (a\xi)^2 (b\eta)^2 J \, d\xi d\eta = a^3 b^3 \int_{-1}^1 \xi^2 \left[\frac{\eta^3}{3}\right]_{-1}^1 d\xi =$$

$$= a^3 b^3 \int_{-1}^1 \frac{2\xi^2}{3} d\xi = \frac{2}{3}a^3 b^3 \left[\frac{\xi^3}{3}\right]_{-1}^1 =$$

$$= \frac{4}{9}a^3 b^3 . \qquad (A.32)$$

A.6 Example: Two-dimensional integration and coordinate transformation: paraboloid

Given is the integral $\int_1^2 \int_1^2 (x^2 + y^2 + 1) dx dy$. Change the variables from $1 \le x \le 2$ and $1 \le y \le 2$ to $-1 \le \xi \le 1$ and $-1 \le \eta \le 1$ and calculate the definite integral. Compare the result with the integration based on the Cartesian coordinates (x, y)

A.6 Solution

The relationship between the Cartesian (x, y) and the natural coordinates (ξ, η) can be derived as:

$$\xi = 2\left(x - \tfrac{3}{2}\right) \ \text{ and } \ \eta = 2\left(y - \tfrac{3}{2}\right) \ \text{ or} \tag{A.33}$$

$$x = \tfrac{1}{2}(\xi + 3) \ \text{ and } \ y = \tfrac{1}{2}(\eta + 3). \tag{A.34}$$

Thus, the transformed integral results finally in:

$$= \int\limits_{\xi=-1}^{+1} \int\limits_{\eta=-1}^{+1} \underbrace{\left(\tfrac{1}{16}\xi^2 + \tfrac{3}{8}\xi + \tfrac{1}{16}\eta^2 + \tfrac{3}{8}\eta + \tfrac{11}{8}\right)}_{g(\xi,\eta)} \mathrm{d}\xi\,\mathrm{d}\eta. \tag{A.35}$$

The analytical integration of Eq. (A.35) gives $\tfrac{17}{3}$.

References

1. A. Holder, J. Eichholz, *An Introduction to Computational Science* (Springer, Cham, 2019)
2. O.C. Zienkiewicz, R.L. Taylor, *The Finite Element Method*, vol. 1. The Basis. (Butterworth-Heinemann, Oxford, 2000)

Index

© The Editor(s) (if applicable) and The Author(s), under exclusive license to Springer
Nature Switzerland AG 2026
A. Öchsner, *An Introduction to the Classical Approximation Methods
in Applied Mechanics*, SpringerBriefs in Computational Mechanics,
https://doi.org/10.1007/978-3-032-06967-2

If you have any concerns about our products,
you can contact us on
ProductSafety@springernature.com

In case Publisher is established outside the EU,
the EU authorized representative is:
Springer Nature Customer Service Center GmbH
Europaplatz 3, 69115 Heidelberg, Germany

Printed by Libri Plureos GmbH
in Hamburg, Germany